Business Agility

Strategies for Gaining Competitive Advantage through Mobile Business Solutions

FINANCIAL TIMES PRENTICE HALL BOOKS

For more information, please go to www.ft-ph.com

Deirdre Breakenridge
Cyberbranding: Brand Building in the Digital Economy

Jonathan Cagan and Craig M. Vogel
*Creating Breakthrough Products: Innovation from Product Planning
to Program Approval*

Subir Chowdhury
The Talent Era: Strategies for Achieving a High Return on Talent

Sherry Cooper
Ride the Wave: Taking Control in a Turbulent Financial Age

James W. Cortada
*21st Century Business: Managing and Working
in the New Digital Economy*

James W. Cortada
Making the Information Society: Experience, Consequences, and Possibilities

Aswath Damodaran
*The Dark Side of Valuation: Valuing Old Tech, New Tech,
and New Economy Companies*

Nicholas D. Evans
*Business Agility: Strategies for Gaining Competitive Advantage
through Mobile Business Solutions*

David Gladstone and Laura Gladstone
*Venture Capital Handbook: An Entrepreneur's Guide to Raising Venture Capital,
Revised and Updated*

David R. Henderson
Joy of Freedom: An Economist's Odyssey

Dale Neef
E-procurement: From Strategy to Implementation

John R. Nofsinger
*Investment Madness: How Psychology Affects Your Investing…
And What to Do About It*

Jonathan Wight
Saving Adam Smith: A Tale of Wealth, Transformation, and Virtue

Yoram J. Wind and Vijay Mahajan, with Robert Gunther
Convergence Marketing: Strategies for Reaching the New Hybrid Consumer

Business Agility

Strategies for Gaining Competitive Advantage through Mobile Business Solutions

Nicholas D. Evans

Prentice Hall PTR
One Lake Street
Upper Saddle River, NJ   07458

Library of Congress Cataloging-in-Publication Data

Evans, Nicholas D.
  Business agility : strategies for gaining competitive advantage through mobile business solutions / Nicholas D. Evans.
    p. cm.
  Includes index.
  ISBN 0-13-186554-4
    1. Business--Communication systems. 2. Mobile communication systems
    3. Telecommunication systems. I. Title

HF5541.T4 E93 2002
658.8'4--dc21

2001052390

Production Editor: *Wil Mara*
Acquisitions Editor: *Jim Boyd*
Editorial Assistant: *Allyson Kloss*
Marketing Manager: *Bryan Gambrel*
Buyer: *Maura Zaldivar*
Manufacturing Manager: *Alexis R. Heydt-Long*
Cover Designer: *Nina Scuderi*
Cover Design Direction: *Jerry Votta*
Art Director: *Gail Cocker-Bogusz*
Composition: *Ronni Bucci*

 © 2002 Prentice Hall PTR
Prentice-Hall, Inc.
Upper Saddle River, NJ 07458

The publisher offers discounts on this book when ordered in bulk quantities. For more information contact: Corporate Sales Department, Prentice Hall PTR, One Lake Street, Upper Saddle River, NJ 07458. Phone: 800-382-3419; Fax: 201-236-7141; E-mail: corp-sales@prenhall.com.

Printed in the United States of America

10 9 8 7 6 5 4 3 2 1

ISBN 0-13-186554-4

Pearson Education LTD.
Pearson Education Australia PTY, Limited
Pearson Education Singapore, Pte. Ltd
Pearson Education North Asia Ltd
Pearson Education Canada, Ltd.
Pearson Educación de Mexico, S.A. de C.V.
Pearson Education—Japan
Pearson Education Malaysia, Pte. Ltd
Pearson Education, Upper Saddle River, New Jersey

Acknowledgments

This book owes a great deal of thanks to the people that have inspired me as leaders in their respective industries and professions. The executives and venture capitalists who have expressed interest and enthusiasm for this project and who helped me with success stories, manuscript review, and in validating my vision for *Business Agility*.

In particular, I'd like to thank (in alphabetical order by company): Mark Sherman (Battery Ventures), Randy Chappel and Craig Thomas (Goff Moore Strategic Partners), and Rob Miles (Vortex Partners).

I'd like to thank Jim Boyd, Executive Editor at Financial Times Prentice Hall, for his interest in bringing my idea to fruition and for our many conversations (often highly amusing) during the course of the project. I'd also like to thank Crissy Statuto for the initial referral to the folks at FTPH, and Wil Mara, Nina Scuderi, and Ronnie Bucci for their great work on the production side.

For their help in providing case studies and company insights, I'd like to thank (in alphabetical order by company): Theresa Enebo and Chuck Grothaus (ADC Telecommunications), Joe Lacik and Jim Park

(Aviall), Scott Heintzeman and Carol Nissen (Carlson Companies), Nancy Liberman and Tom Svrcek (EnvoyWorldWide), Susan Burud (FT Interactive Data), Laura Rippy (Handango), Petri Karppinen (More Magic), Dennis Andruskiewicz, Kevin Conklin and Lauren Garvey (Office Depot), Simon Gawne (Red-M), Jeff Cummings (Rental Service Corporation), Amir Alon, Nick Howarth and Nick Ward (Sirenic), Paul Reddick (Sprint PCS), and many others. Thank you all for taking the time to talk with me and for sharing your thoughts and successes.

Finally, I'd like to thank my wife, Michele, and my two sons, Andrew and David, for their patience with me and for allowing me to take family time on evenings and weekends in order to put this book together. Without your encouragement and support, this book would not have been possible.

Readers who would like to correspond with the author can contact him at nick@nevans.com.

Contents

2 The M-Business Evolution   27

3 Design of an M-Business   69

4 Process Models and Applications for M-Business Agility   85

5 Industry Examples   119

6 M-Business Strategic Roadmap · 151

7 M-Business Architectural Frameworks · 169

Foreword

There is little doubt that recent developments in the wireless and information technology industries will have a profound influence on the way business is done. Beyond that, the new "way" is still emerging, and the path to get there is still unclear. These new technologies only have value to the extent that business processes, in their broadest sense, take advantage of the new capabilities they provide. The challenge is to explore the interaction between new technologies and business processes in search of insights that can guide businesses to the new "way."

Nick Evans is an intellectual pioneer of sorts. He did not invent M-Business any more than Lewis and Clark invented the Rocky Mountains. However, by observing more closely than most what is transpiring in the early days of M-Business, he details what he sees and extracts key learnings that are valuable to companies navigating through the various stages of their own M-Business strategies. Consider his current writings as letters—an early glimpse at both the opportunities and challenges that lie ahead.

Over the past several decades, businesses have become increasingly aware of the importance of understanding and improving the processes that result in increased productivity and greater competitive advantage. This understanding of the effectiveness of business processes and its correlation to an organization's goals (e.g., profitability) drive business change. This is not an entirely new phenomenon, just one that has accelerated.

Changes in information technology are accelerating even faster. In a relative nanosecond of history, we have seen the rapid evolution of mainframes, PCs, client-server architectures, the Internet, and the growth of distributed computing.

Today, advancing wireless technologies are now serving as a tremendous catalyst. If the need to improve business processes is the fuel, and the developments in information technology are the oxygen, then advances in wireless capabilities are the sparks for incredible change. How an enterprise conducts its business, how it relates to its customers, and how it works with its partners will all be affected by this inevitable change.

So enough about the hype. Beyond the general excitement and a rough understanding of the possibilities, what can we do today to ensure that we are neither left on the backside of a wave nor teased into the break? A few thoughts:

1. Have an M-Business strategy. Even if you choose to go it slow, do it consciously.

2. Think big. An M-Business strategy is more than extending corporate e-mail to wireless devices.

3. Pursue the principle of perfect and ubiquitous information. Ask what kind of things could happen if an employee, a customer, or a partner had the right information, at the right time, in the right place, and/or the ability to act on that information ...then ponder why they do not have it.

4. Know the key technologies. Failure to work around their weaknesses or to leverage their strengths can be the difference between success and failure.

5. Clearly quantify the expected and experienced impact of improvements. Real waves worth catching have real ROIs.

6. Do not underestimate the implementation challenges, especially related to human factors such as training. Few things are more frustrating to watch than employees or customers who put themselves through unnecessary punishment by doing things the "old way."

7. Be comprehensive when considering solutions with a bias towards open standards. Proprietary solutions are especially vulnerable given the multiple moving parts (e.g., devices, client software, networks, network solutions, and internal systems) that must all work in harmony.

8. Experiment and learn. Do not wait for your competition to teach you.

In this new environment, there is a lot of opportunity, no shortage of options, and a lot to consider. Making it happen is no small order. This book grants an opportunity to skip part of the learning curve by defining important concepts and frameworks, putting into context the key technologies, players and market forces, identifying opportunities for creating enterprise value, and providing a window to observe the results of others' experiments. My hope, and I believe that of the author's, is that those who explore early and intelligently in this space will prosper.

Paul Reddick
VP Business Development
Sprint PCS

1

Introduction

Mobile Business = Business Process + Electronic Business + Wireless Communications

Mobile business, or M-Business, is already changing our lives. It is often thought of as the business opportunity that has been brought about by the convergence of electronic business with wireless technologies. We can now conduct business anytime and anyplace in order to meet the informational and transactional demands of end users, remove former process and technology bottlenecks, and hence increase customer satisfaction, revenues and productivity, and reduce costs.

But beyond this simple description, we are entering an entirely new era—the era of business agility, which is the next evolution of business itself. This is a superset, not a mere extension, of electronic business. It is here that mobile business will have its most profound effect. The changes will be widespread. They will affect our companies, our markets, our economy, our lifestyles, our politics, and our global perspectives and interactions.

This book is designed to serve as a handbook for this next business evolution. The era of Business Agility presents us with a new era of enterprise value and productivity after the crash and disappointment of the prior wave of E-Business initiatives. Over the past several years, businesses have made great strides in improving customer service, productivity, and manufacturing processes. Businesses have optimized their relatively unchanging processes and value propositions and have gained considerable efficiencies. Information technology has played a large role in facilitating these increases in productivity. But the next competitive frontier will be how quickly these businesses can sense and respond to change. The next step is to be able to create new processes and new value propositions in a rapidly changing business environment and be able to institute these changes rapidly. These changes must also have a higher degree of flexibility engineered-in once implemented. In an era when entire industries are converging, enterprise agility, and the ability to create dynamic value propositions, will separate the winners from the losers in the battle for mindshare and marketshare.

The ability to leverage M-Business to pull information rapidly and accurately from the "edge" of the corporation in the areas of sales, marketing, and customer service will be critical for the agile enterprise. Likewise, the ability to leverage M-Business to push management decisions back out to the field, where those decisions can be acted upon, will be critical as well. M-Business represents an opportunity to increase business agility across the entire breadth of the corporation—from its core to its extremities—connecting management with employees, customers, suppliers, and business partners.

Without this business agility, even companies that have highly optimized processes and excellent customer service today will find themselves struggling to compete over the next several years owing to information and transaction chokepoints that delay decision making and employee action. Not only will they experience a critical time delay in their enterprise operations and customer and supplier interactions, they will also experience information gaps where they are less informed of critical business events and metrics in their surrounding environment. They will be less aware of the business context in which they operate and will lack a competitive edge.

Figure 1–1 M-Business Viewed as a Superset of E-Business.

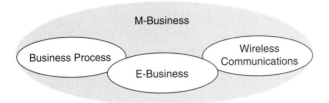

M-Business presents a new opportunity to enhance business agility. Figure 1–1 shows the broader view of M-Business that is the subject of this book. Rather than being at the intersection of E-Business (or the Internet) and Wireless Communications, which has been the traditional definition of the wireless Internet, it is a superset comprised of business process, electronic business, and wireless communications. It is as much about process as it is about technology. The combination of all of these areas creates powerful new synergies.

Throughout the course of this book, we'll take a look at the formula for business agility and present six principles and nine action items for achieving business agility within your organization. We'll look at the current state of affairs with regard to the wireless industry, the drivers and barriers to adoption, the wireless Internet value chain, and the applications that are gaining traction in the industry. We'll look at concepts such as the dynamic value proposition, how to leverage preference-driven commerce within your customer interactions, and how to apply alliance relationship management within your partner interactions.

We'll explore areas of opportunity within the enterprise for M-Business to improve business intelligence, customer relationship management, sales force automation, field force automation, and supply chain management.

We'll explore case studies from companies such as ADC Telecommunications, Aviall, Carlson Hospitality, eBay, FT Interactive Data, Office Depot, and Rental Service Corporation in order to see how these early adopters leveraged M-Business within their enterprise and achieved real business results.

We'll look at M-Business strategy and how to calculate return on investment. We'll look at M-Business architecture and implementation, and profile some of the key wireless software vendors, in order to execute on the strategy.

Finally, we'll look at future trends in M-Business and other emerging technologies such as the semantic Web, real-time computing, Web services, natural language processing, and business process management. All these emerging technologies are helping us adapt technology around ourselves and our business activities—enabling us to increase our business agility.

Formula for Business Agility

As we noted at the beginning of this chapter, the following formula can help explain the extent and value of mobile business:

Mobile Business = Business Process + Electronic Business + Wireless Communications

Business constantly seeks ways to improve itself, whether it has the goal of increasing shareholder value, increasing customer satisfaction, increasing revenues, or reducing costs. Over the years, we have seen several trends develop. Some of the most recent trends have been business process re-engineering (BPR), enterprise resource planning (ERP), electronic business (E-Business), and the growth of wireless communications. If we view these trends by themselves, they have been evolutionary steps. What is fascinating is that we are now entering an era where emerging technology is allowing us to combine them in powerful new ways that signal what could well be a true business revolution—that of Business Agility. M-Business is one of the first waves coming to shore that signal this impending change in the landscape of possibilities.

The following formula can help explain the requirements for business agility:

Business Agility = Process Agility + Technical Agility

M-Business is the fuel for achieving Business Agility because it carries with it both process agility and technical agility. M-Business is not simply E-Business unwired or untethered. Nor is it simply another channel for E-Business. It's a whole new opportunity to create new business processes and improve old ones, while at the same time leveraging the time and location sensitivity that wireless communications brings to an enterprise. Successful M-Business is not just about technological innovation—it requires an equal part of process innovation.

It is important to note that we must also consider the importance of human agility. Throughout both the process agility and the technical agility components of our formula for business agility, the human element is paramount. While I have not included human agility in the formula, please consider it an essential component that permeates across everything we discuss.

Principles of Business Agility

The following principles describe the forces that are driving us toward the concept of Business Agility as the next business revolution:

Principle #1: The Digital Economy Demands Business Agility

Today's global economy and market conditions require that companies in all industries optimize their performance in all areas. The requirement for continuous improvement drives the need for business agility. A company that can sense and react to changes in its internal and external environment more quickly than another can seize new revenue opportunities and control costs more efficiently than its competitors. First to market may not always be attractive, but first to react, based upon sound judgement, is always a positive.

For an organization to increase its business agility, it needs more timely access to information upon which to base decisions and more timely mechanisms for executing those decisions. It needs to have more efficient linkages with employees, customers, suppliers, partners, stakeholders, and even competitors. Business agility needs to be holistic in scope—all touchpoints that have a bearing on the business need to be addressed.

Business agility means not only fast reaction times, but also fast reaction to change and the ability to rapidly implement change. Thus, we have the formula:

Business Agility = Speed x Flexibility

To provide some examples of business agility, we'll take a look at a company's relationship with its customers, suppliers, partners, and competitors (Table 1–1). Business agility within this framework can be illustrated as follows:

Using the Seven "S" model (Figure 1–2), we can look at business agility from the perspective of the fabric and overall structure of an organization. The Seven "S" model was developed by Thomas J. Peters, Robert H. Waterman, and Julien R. Phillips. It is a popular way to look at strategy as part of an organization.

In this context, business agility means having speed and flexibility ingrained into corporate culture and style. This is a top-down approach,

Figure 1–2   The Seven "S" Model.

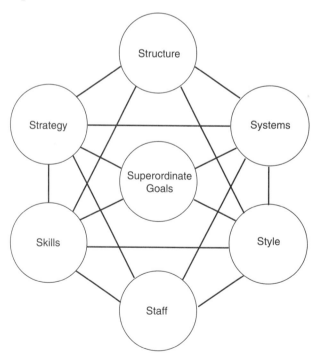

Table 1–1 Examples of Business Agility Applied to Customers, Suppliers, Partners, and Competitors

Customers

Speed

>> How quickly can you acquire customers, deliver services and products, service and support these customers, and capture their feedback?

Flexibility

>> How flexible are you in adapting to customer preferences and needs and in modifying your sales and marketing strategy and product/service mix?

>> Can you provide dynamic "offers" and products and services that respond to real-time customer demands?

Suppliers

Speed

>> How quickly can you communicate with suppliers, manage supplier relationships, manage quality, and act on this information?

Flexibility

>> How flexible are you in adapting to supply chain fluctuations?

>> Can you dynamically adjust your supply chain processes and partners?

Partners

Speed

>> How quickly can you communicate with partners, manage partner relationships, manage quality, and act on this information?

Flexibility

>> How flexible are you in your ability to form and disband partnerships based upon enterprise requirements?

Competitors

Speed

>> How quickly can you capture competitive intelligence data and act upon it?

Flexibility

>> How flexible are you in adapting to new entrants and changes in the competitive landscape?

>> Can you dynamically adjust your competitive positioning?

with the concept of business agility as a super-ordinate goal that forms the mission of the company. As such, it is ingrained in the structure of the organization, in the corporate, business and functional strategies of the organization, in IT systems and business processes, in human resources systems and processes, and finally in the skill sets of the employees themselves.

When business agility is ingrained into every aspect of the corporation, there are no weak links. Each of the Seven "S"'s interoperate in harmony.

Of course, business agility is not just about speed. It's about doing purposeful work in alignment with corporate objectives. Doing this work in an agile new manner enables a quicker response in terms of speed and flexibility.

Principle #2: Business Agility Involves Shaping Technology Around Ourselves

Business agility relates to the speed and flexibility of an organization, but it can also be a term that is used to describe some of the broader trends going on within the technological world at the present time. Business agility is a desirable end-state for the enterprise, but it also helps to describe where technology is heading over the next several years.

Emerging technologies are allowing us to start to shape technology around ourselves. For a long time, we had to work in the other direction, i.e., shape ourselves around rigid IT applications and processes. Today, technology is gaining more human-like characteristics. There are wearable computers with head-mounted displays, voice interfaces such as those found in new vehicle telematics systems, intelligent agents for helping us sort through vast amounts of content and identify the most relevant items, and semantic Web applications that add meaning to Web content for computers to better understand. This new breed of enabling technology will allow us to communicate more naturally with computers. Computers will soon be able to "follow" us. They will be more readily accessible, know our location and our environmental context and activities, and will better "understand" our meaning and objectives. By adapting around us in this way, these emerging technologies are delivering higher levels of value in our daily lives, both personally and professionally. They are also becoming less invasive in terms of how and when we interact with them.

Principle #3: Business Agility Is Achieved Via M-Business

As a critical enabler for these emerging technologies, M-Business applications have unique characteristics that are required for business agility. Most particularly, they have the ability to deliver information at any time and any place, thus reducing the limitations of time and space, and some of the barriers that we encounter when using the wired Internet.

In fact, M-Business presents us with a unique technological era, where we can adapt and shape technology to ourselves as opposed to adapting ourselves around technology. As human beings our day-to-day work activities are more about process (i.e., task-oriented) than about technology. Thus, as technology starts to adapt to fit around us, it naturally is shaped more and more by process.

M-Business can be the strategic enabler of business agility across all of the enterprise functional areas from sales and marketing, finance, human resources, operations, business development, product development, and support and services to the IT department itself.

The golden rule here is to always remember that M-Business is the combination of both process and technology. As such, it needs to mold itself to fit the business need for information and transactions at the point of business.

Principle #4: Every Business Will Become an M-Business

Businesses adopt technology and process innovations if they are deemed to add some measure of value to the enterprise. Over the years, businesses have adopted telephones, copiers, fax machines, mainframes, client-server technologies, Internet technologies, and many other devices and technologies in order to improve communications and manage information. The technology adoption lifecycle shows how technology is adopted at various stages of its evolution by innovators, early adopters, the early majority, the late majority, and laggards. These stages of technology adoption show which enterprises are most aggressive in their application of technology for business advantage and which are most conservative. Timeframes across the adoption lifecycle vary considerably based upon the technology in question. Figure 1–3 shows the technology adoption lifecycle.

Figure 1–3   Technology Adoption Lifecycle.

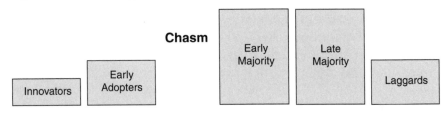

It is fair to say that M-Business applications for the enterprise are currently in early adoption status. Supply of technology solutions currently far outweighs demand in the enterprise, mostly because the business benefits have not been articulated and the business value not well proven or understood.

Figure 1–4 shows the adoption rates of some of the technologies mentioned above. In comparison, the current adoption rate in the mobile device market has reached in excess of seventy percent in some countries within the space of just a few decades.

Figure 1–4   Various Product Adoption Rates.

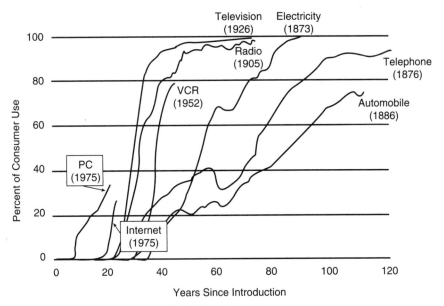

Despite the early adoption status of M-Business applications for the enterprise, it is clear that the device penetration and adoption of mobile communications has outstripped even the growth of the Internet itself. With this momentum helping to lay the infrastructure for M-Business, it is evident that the pervasiveness of M-Business software and applications will not be far behind.

Along with the technology adoption process comes the technology hype cycle as coined by Gartner, a research and advisory firm. This hype cycle explains the maturity of new technologies as they come to market. Initially, as a "technology trigger," they generate a lot of hype leading to the "peak of inflated expectations." As real-life implementations discover the weaknesses in addition to the strengths of the technology, the technology falls into the "trough of disillusionment." Finally, as enterprises learn to gain productive usage from the technology, it emerges from the trough and reaches the "slope of enlightenment" and then finally the "plateau of productivity." This technology hype cycle can equally be applied to E-Business or M-Business in general as well as to specific technologies.

The key takeaway is that new technologies will almost always be over-hyped and lead to a period of disillusionment before productive use is finally extracted. This is really a difference in the observed value of the technology when compared to the expected value at the specific point in time.

Having looked at the technology adoption lifecycle, the product adoption rates, and the technology hype cycle, we can now take a look at a macro-perspective of where business agility and the technology for helping us achieve this agility is heading. By looking at business agility from a process and technology perspective, Figure 1–6 shows the evolution since the mid-90's when Internet applications were first unleashed as a true business driver.

A four-phase model of this evolution is as follows:

>> **Phase I—Internet**—This phase was characterized by static Web pages that allowed companies to present product information to customers, company information to investors, and so on. It started the process of focusing the IT department externally beyond corporate boundaries in addition to its typical internal focus. Admittedly, electronic data interchange (EDI) applications

Figure 1–5 Hype Cycle of Emerging Technology. Source: Gartner Website Glossary, July 2001.

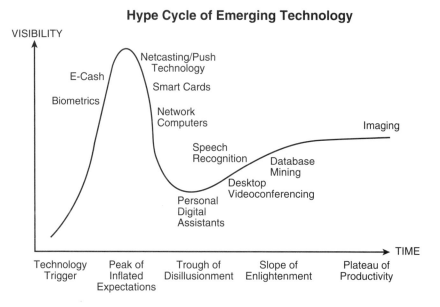

Hype Cycle of Emerging Technology

and other B2B processes had been in existence prior to the Internet, but the Internet was the first time IT had found a mass market of consumers and business partners for which to provide their services. Internet Web pages provided some technical agility for the enterprise, but little process agility.

>> **Phase II—E-Business**—The E-Business phase was characterized by dynamic business applications instead of static Web pages. These applications were leveraged in business-to-consumer, business-to-business, and business-to-employee scenarios. These dynamic applications made it possible for true transactions to be conducted online with data driven by corporate databases. E-Business applications provide greater technical agility for the enterprise and also started to add process agility as phone, fax, and paper-based processes were re-designed and often eliminated in order to take advantage of E-Business processes. The beginnings of collaborative commerce added to the level of process agility at this stage.

>> **Phase III—M-Business**—The M-Business phase leverages the time- and location- advantages of mobility in order to increase the process and technical agility of the enterprise. This is the current state of the evolution of business and we are just at the beginning. The M-Business era has garnered a lot of attention in the press, but it's important to note that E-Business technologies are also maturing and continuing to evolve alongside mobile technologies. One example is the semantic Web, which aims to add meaning to the Internet as it exists today in order to enable computers to better understand the meaning of our data and to make intelligent decisions based upon that meaning. Other examples include real-time computing, natural language processing, Web services, and business process management—all of which are discussed in more detail in our Future Trends chapter later on.

>> **Phase IV—I-Business**—This can be considered as the theoretical end-state beyond M-Business, where companies are on an equal footing in terms of their leverage of technology and are forced to compete solely on intellect and corporate strategy. The "i" can be considered as standing for intellect, ideas, and innovation. Many companies at this state may have totally outsourced all their physical production and exist solely as a brand owner. This end-state can be considered as a virtual game of chess. The corporation is entirely virtual and, yes, business decisions can be executed at the speed of thought.

In summary, every business will become an M-Business. This simply takes time. The ability to pull information and transactions from "edge" employees in areas of sales, marketing, and customer service and to push decisions out from management to these employees to act upon is too compelling to ignore.

What needs to happen is that the technology needs to move through the technology adoption lifecycle and into the mainstream enterprise adoption. The various technology triggers that enable M-Business need to mature through the hype and disillusionment phases that we all know too well and move into the phase of true productivity, where enterprise value can be extracted. Additionally, the enterprise will require solid execution in addition to this new technology in order to be successful and to achieve its' business agility goals.

Figure 1–6   Business Agility as a Function of Process and Technology Agility.

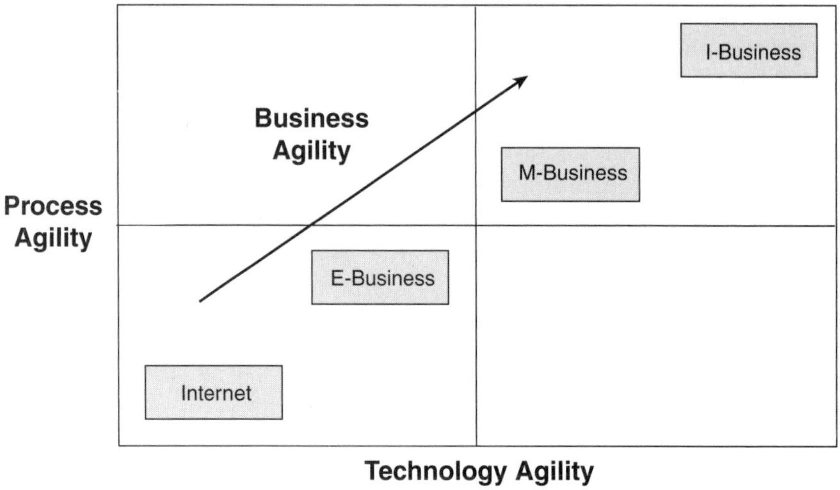

Principle #5: M-Business Will Drive Both Business and Technical Transformation

As soon as technology enabled business to capture increased revenues or reduce costs, it started to have a loop-back effect. Not only could the business side of the enterprise create a new process to better serve customers or to increase productivity—and request and hence drive a technical transformation—but the IT side of the enterprise could also create a new process via a new technology and thus drive a business transformation from the other direction.

Technology has been benefiting business for hundreds of years, but only recently has it been able to make such profound changes possible in such short timeframes. A new mechanical process for manufacturing twenty years ago would create competitive advantage, but would take a long time to implement and had high asset intensity in terms of the capital expenditure involved. Today's software innovations can transform a business process within weeks or even days and have an immediate impact on the business. One of the drawbacks, however, is that these innovations can be more easily copied by competitors.

As we start to explore this new environment where information and transactions are now time and location aware—where users are mobile and devices are portable, the M-Business era can drive both business change and technology change. In this fashion, it is self-propelling, feeding its own transformation.

Principle #6: Industry Convergence Creates New Threats and Opportunities

Within the space of a few short years, industry convergence has moved from an interesting theoretical end state on a whiteboard discussion of electronic business into a real-life day-to-day phenomena.

It is affecting almost every industry imaginable. Not only are the computing, communications, and media and entertainment industries converging, but we are also seeing the travel and transportation industries becoming information providers, banks becoming retailers, railroad companies becoming phone companies, real estate and hospitality companies becoming communications providers, energy companies becoming business-to-business marketplaces for bandwidth, and many other examples.

This industry convergence means that not only will your enterprise experience competition from new entrants to your traditional industry, but that your industry will find itself transformed into a next generation industry with new business models and new rules. Competing effectively in this new industry will require strong business agility in order to re-position for new markets, new customers, new products and services, new competitors, and new regulations.

Your position in the value chain may be affected as well. Without taking action you may find your products and services pushed further from the end customer as new players create new business models with new value chains. Conversely, convergence may create new opportunities to become a key player in a new value chain. For example, as wireless carriers start to provide data services to the enterprise and provide M-Commerce services for consumers, opportunities are created for companies to play the role of transaction enablers with financial services, security services, preference services, location-based services, and advertising services to name just a few.

Action Items for Business Agility

The following action items are steps that your enterprise can take in order to move toward business agility. These are the high-level steps that will be elaborated upon throughout the course of this book. The goal is to drive a change in process, technology, and culture. Such a change will best prepare and enable the enterprise to rapidly sense and react to changes in its environment. To do so, an enterprise must attempt to move information and transactions closer to the point of business where they are required, and to remove existing technology and process bottlenecks in the delivery of this information and transactions. This requires new process, new technology, and new attitudes.

Step #1: Make M-Business Part of Your Business Strategy

Firstly, M-Business needs to be part of the overall business strategy. Ensure that your functional, business, and corporate strategies all take M-Business into consideration. At the functional strategy level, M-Business can be applied to improve your operating methods. At the business strategy level, it can help you fight your competition in your current industry. Finally, at the corporate strategy level, it can help you re-think the entire suite of business opportunities available and potentially enable you to enter entirely new industries.

A simple example is a country club that enables its premises with a wireless LAN technology such as Bluetooth. In this way, it is providing new, differentiating services to its members and could be considered as entering the data services industry by allowing members to check e-mail and their corporate intranet via their PDAs while at the country club.

Another example is a credit and transaction services company that provides private-label credit cards to the retail, petroleum/convenience stores, utility, and transportation industries. By making M-Business part of their business strategy, they have an opportunity to provide financial services to the wireless carriers who need to bill for data services and content services provided by their partners over their wireless network. Since the carriers often do not wish to become financial institutions, the credit and transaction services company has the opportunity to provide these services for the wireless carrier on behalf of their content providers. Additionally, their extensive loyalty and database marketing information from their existing customer base can be leveraged for greater customer personalization when conducting wireless transactions—something the wireless carriers may not have.

Step #2: Make the IT Department a Strategic Partner

To achieve business agility within your enterprise, one of the vital first steps is to make the Information Technology department a strategic partner of the overall business.

Over the past five years, the IT department has rapidly evolved from being a cost center and support function, to a business enabler, and finally to a legitimate business entity unto itself.

As a cost center and support function, IT responded to requests from the business to provide custom reports and to build customized applications in order to capture and process data. Even as recently as the mid-90's, this was still the mindset and function of corporate IT. Client-server applications provided just this function and with their leverage of Windows graphical user interfaces they began to make IT applications usable for end users and not just IT staffers.

As a business enabler, IT implemented ERP systems in the mid-90's and built custom Web applications in the late 90's. These Web applications started to reach out from within the enterprise to customers and business partners via the Internet and with extranets. IT also simplified information management within the enterprise via corporate intranets. Thus, the wave of the Internet was the catalyst that extended the reach of IT out to customers, suppliers, and business partners. This was happening to some extent in the client-server days, but the deployment and maintenance of applications to external constituencies was laborious in terms of both time and cost.

Finally, with the wave of E-Business, we saw IT become a business unto itself. Internet-related IPOs outpaced non-Internet IPOs in the United States for a period of five consecutive quarters between the first quarter of 1999 and the first quarter of 2000, as shown in Figure 1–7. While the bubble has since burst, the main takeaway here is that information technology can become a business unto itself and a significant contributor to enterprise value.

As we move into the next era of M-Business, IT will become an even more critical business partner owing to the convergence of the computing and communications industries and their ability to create competitive advantage for the enterprise. IT has already made significant progress toward being a strategic partner. In fact, in an online poll conducted by Internet Week, 86% of respondents stated that they believed both IT and business management should be involved in IT decisions—confirming the strategic nature of those decisions.

Step #3: Pursue a Holistic M-Business Strategy

It is important to pursue a holistic M-Business strategy that takes every aspect of information delivery and transactions into account. The wireless and wireline portions of application delivery need to be seamless and coordinated. If you are not familiar with the term "wireline," it simply means "wired." It's a term that is often used in the telecommunications industry to differentiate between the wireless phone service and the wired phone service sides of the business.

Figure 1–7   Internet Versus Non-Internet IPOs. Source: VentureOne.

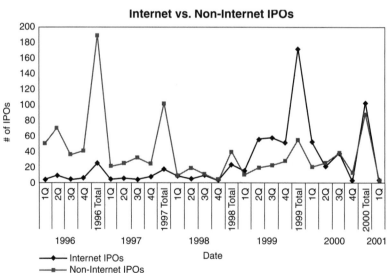

The wireless aspects need to extend the relationship with the customer, giving the customer more options for access and information, but need to be framed as an additional entry point into the enterprise relationship. Information and transactions need to follow the end user no matter how he or she is accessing the system—whether over wireline connections such as kiosk, phone, fax or PC, or over wireless connections such as PDA, 2-way pager, or WAP phone.

On a technical level, this means that a holistic approach needs to be taken in order to create business applications that follow the user across these devices and have a consistency in terms of security and session management. Ideally, a single access and authentication technique such as a username and password enables the end user to access your systems via any device. His or her transaction should also be able to be started on one device, paused, and then continued on another device as needed. This is termed session management—the users' session or context moves with the user.

In addition to the holistic view of access to information and transactions via any device and any communications method, the view of the user constituencies should also be holistic. Often, information and transactions need to be made available not only to customers, but to employees, partners, and suppliers. Various levels of security can be used to restrict the views of the data based upon user role, but the data and applications need to be consistent.

Step #4: Exploit and Defend Your Position in the New M-Business Value Chain

It is important to understand the new value chains that are being created and those that can be created. By understanding the current M-Business value chains that are being formulated and the realm of possibilities for future value chain configurations, you can best plan your next move. Ask what role your company might be able to play in these new value chains. How easy would it be to enter and defend this position? Who are the other entrants? Do they have the core competency and the brand recognition to accomplish the task? How can your companies' core competencies be leveraged in these new areas? What are the needs of your customers? How do they view these new value chains and how do they extract value from the products or services delivered? What alliances do you have that can be leveraged to create an entirely new value proposition for your customers?

Understanding your position in the new value chain means studying not only your traditional competitors and your customer needs, but also looking at industry convergence and determining how these converging industries are changing and expanding their value proposition to your customers. If they succeed, are they likely to own the customer more than you will? Are they exploiting the time and location sensitivity of mobile business in order to reach out to your customers when they are on your premises?

An example is in the travel industry. A business traveler sees a business trip as a single event that needs to be planned from beginning to end. But during this trip many service providers are involved: travel agencies, airline companies, rental car companies, hotels, employers (from the time and expense perspective), credit card companies, and so forth. The company that "owns" the customer may end up being the company that provides the most useful service during the entire lifetime of the trip from planning, through to the trip itself, and finally to the time and expense reporting after the trip. By leveraging mobile business technologies such as a WAP phone or PDA, the resourceful full-service provider can communicate with your customer when he or she is in your airport, your hotel, or your rental car. This relegates you to a role where the full-service provider and "owner" of your customer can potentially advise the customer to travel out of your travel network and take a portion of his or her trip with your competitor.

Another example is in the retail industry. A comparison shopping service delivered over mobile devices can tell a customer in your store where a better price is down the road at a competitor's store. Today, companies such as BarPoint provide comparison shopping services for books, movies, music, computers, and electronics accessible to users of PDAs and WAP-enabled cell phones. Users simply enter UPC numbers from the barcodes on the products they are interested in and a list of stores carrying that product is returned and displayed on the device. It will not be long before these types of services add location-based functionality to make the service even more useful for consumers so that they are pointed to the "better deal" down the street.

The examples above should serve to illustrate that it is vital to both exploit your position in the new value chain, but also to seek ways to defend your position in the existing value chain as well.

Step #5: Design Business Processes to Take Advantage of M-Business

In meeting the business need for information and transactions closer to the point of business, there are two major considerations around business process. Firstly, can an existing business process be improved and streamlined through the use of mobile business technologies? Secondly, do mobile business technologies present entirely new opportunities for new business processes in addition to the existing processes?

As an example, take the case of the field service worker employed by a telephone company. Even if this person uses a laptop with a wireless modem in his or her truck in order to pull up work orders, can this existing process be improved via mobile business technologies? If the worker is going into the customer premise and making notes that are later transferred to the laptop in the truck, the answer is most likely yes. By taking a PDA or other device to the actual point of business, in this case the customer premise, the worker can avoid the double entry involved in making hand-written notes and then transferring them to the laptop. This process improves the existing process by moving the information and transaction functionality closer to the actual point of business.

Additionally, new business opportunities for this mobile worker may include the ability to cross-sell additional services to the customer by accessing product and pricing information over the PDA.

Step #6: Design Technical Architectures to Take Advantage of M-Business

In order to take advantage of the new business opportunities presented by mobile business and to meet the business need for information and transactions closer to the point of business, it is also very important to re-think and re-design IT technical architectures. In addition to a holistic M-Business strategy, a holistic IT strategy is also required. Just as CEOs mandated a Web presence for the company a few years ago during the Internet and E-Business era, CEOs will soon be mandating a mobile presence. Even conservative companies will recognize the opportunities for business agility inherent in mobile business technology and will plan to support this new channel for information and transaction flow.

To support these business drivers, the CIO and the IT organization will need to plan to further open up the enterprise to mobile devices and mobile users. This is a daunting task given the nascent state of the mobile industry at the present time. However, it is important for IT to realize that these requirements will be upon them sooner than expected and that a strategy and action plan needs to be created for the transition of IT architecture and applications to the mobile world.

If it has not done so already, IT needs to immediately start planning its mobile business strategy and start to research all aspects of how it will extend its relevant applications to the wireless arena. The IT strategy should encompass a detailed look at the following assets, services (including vendors), and skills: wireless devices, user interfaces and browsers, wireless standards, protocols and best practices, wireless security (including access control, user authentication, and encryption), wireless development environments, wireless application middleware, wireless extensions to packaged application, wireless carriers and service providers, service level agreements, maintenance and support of wireless applications, and plans for outsourcing versus insourcing of applications and services.

Step #7: Design for Rapid Change in Process and Technology

When implementing M-Business initiatives, be sure to design for rapid change. From an end user perspective, this means allowing the end user to constantly provide input for improvements and suggestions for new functionality. The applications should be allowed to evolve over time and take on new forms. From the IT perspective, a modular, component-based architecture should be employed with loose coupling between components. The architecture should allow for components to be plugged into the business logic layer and new modules to be plugged into the user interface layer. A portal concept for the user interface layer will help to allow multiple applications to be accessed from the same launch area on the user's cell phone or PDA.

Rapid prototyping and pilots with small groups of end users will help to validate new applications and evolve them based upon feedback from the initial group. This will also help to minimize risk by testing with a subset of the end user population prior to full deployment. The prototypes can be used to test both the slice of business functionality and also the end-to-end functionality from a technical perspective.

To enable rapid change in process and technology, it is also important to ensure that the applications you deploy are well documented. Ensure that the end users have well documented instructions on how to configure their devices, launch their applications, use the applications, and ask for help where needed. The IT department should also ensure good documentation of code and of the overall technical architecture. This is something that is often overlooked within the IT department owing to time constraints, but in order to build applications that can be changed rapidly it is important to have an understanding of the current state.

Step #8: Focus on User Acceptance and Training

One of the most important aspects of a successful mobile business strategy and rollout is user acceptance. The applications and initiatives may have good potential for return on investment, and be well planned architecturally, but without end user acceptance and adoption these initiatives can achieve far less than was originally intended.

End users need to have an input into the requirements gathering process and the functionality that the applications are enabling. Whether they are employees, customers, suppliers, or partners, they are the end customer and need to have applications that are of benefit to them in their daily activities. If the end users are external to the enterprise, the goal is to make it easier for them to do business with you than with your competitors—or to enable them to do even more business with you in a given period of time. If the end users are employees, the goal is to make them more productive.

End users may need help in training on how to use their business applications. User interface design and navigation through the applications is especially important. The small form factors of today's mobile devices mean that information and navigation needs to be crisp and concise. Studies have shown that the number of users accessing applications on WAP devices drops by 50% on every screen. Clearly, key applications and processes need to be enabled with as few navigation steps as possible. Ordering of the menu items and lists needs to also be carefully considered.

User training needs to also focus on the devices themselves. They are currently hard to configure and manage. They need constant recharging due to short battery life, and configuration of wireless connectivity is often hard even for experienced IT practitioners.

What made Internet and E-Business applications so easy to use was their simplicity over the prior generation of client/server applications which had to be installed and configured on each desktop. IT needs to ensure that the ease of use of E-Business applications is translated into an equivalent ease of use for M-Business applications. Without this ease of use and without proper end user acceptance and training, M-Business applications will fall short of their tremendous potential.

Step #9: Measure the Results of M-Business Initiatives

Finally, once M-Business initiatives have been put into place, it is important to track and measure the results of these initiatives and to feed the results back into your overall strategy. M-Business initiatives need to be measured both in terms of the hard benefits and soft benefits. When we speak of hard benefits, we're referring to revenue generation, cost reduction, and increases in productivity and reductions in cycle times. By soft benefits, we mean improved employee moral, increased customer satisfaction, expanded range of options for future business and technology directions, and so forth.

When measuring the results of your M-Business initiatives, it is important to capture metrics at both the business and technical level. The business metrics can be used as described above in order to help shape future strategy and application initiatives. The technical metrics can be used for input back into the IT organization in order to better understand usage patterns and device and application provisioning and support behavior and actual service levels in terms of scalability, performance, availability, and reliability.

Business Agility Lessons

Formulae

Mobile Business = Business Process + Electronic Business + Wireless Communications
Business Agility = Process Agility + Technical Agility
Business Agility = Speed x Flexibility

Principles of Business Agility

Principle #1: The digital economy demands business agility
Principle #2: Business agility involves shaping technology around
ourselves
Principle #3: Business agility is achieved via M-Business
Principle #4: Every business will become an M-Business
Principle #5: M-Business will drive both business and technical
transformation
Principle #6: Industry convergence creates new threats and opportunities

Action Items for Business Agility

Step #1: Make M-Business part of your business strategy
Step #2: Make the IT department a strategic partner
Step #3: Pursue a holistic M-Business strategy
Step #4: Exploit and defend your position in the new M-Business value
chain
Step #5: Design business processes to take advantage of M-Business
Step #6: Design technical architectures to take advantage of M-Business
Step #7: Design for rapid change in process and technology
Step #8: Focus on user acceptance and training
Step #9: Measure the results of M-Business initiatives

2

The M-Business Evolution

In this chapter, we cover the current state of the union within the mobile business community. This will provide a useful backdrop as we build from this starting point, the "as-is" situation, and explore enterprise strategies, case studies, and tactical action plans throughout the remainder of the book. It is important to note that the "as-is" situation has been driven in most part by the supply side of the equation: the wireless handset manufacturers, the wireless carriers, and the wireless infrastructure and software providers—i.e., those who stand to benefit the most from the market creation and adoption. The demand side has picked up mostly in unexpected consumer application areas such as text messaging and gaming. Meanwhile, mainstream enterprise patiently observes in the wings.

Mainstream enterprise adoption is most likely to occur in areas that provide strong business benefits and return on investment. Enterprise success stories around M-Business are appearing more and more frequently. Several success stories from early adopters of M-Business are covered in the chapters on Applications and Process Models for M-Business Agility and Industry Examples. Typical enterprise applications of M-Business have been within wireless enablement of employees: sales force automation and field force automation being two of the most prominent areas with strong returns on investment coming to light.

Of course, the current "as-is" state within the mobile business community is continuously in flux and continuously redefining itself. Analyst predictions for the growth of the wireless Internet and for M-Commerce are merely just that—predictions. They also vary widely between different analyst groups. But despite these variations in analyst predictions, we can still determine clear trends and plan our enterprise strategies accordingly.

Although markets can come and go, and in some cases never meet expectations in terms of potential size and ubiquity of products and services, the convergence of electronic business with telecommunications and other industries such as media, entertainment, and financial services will continue. The drivers toward adoption will become more powerful when compared to the barriers preventing adoption. As the market matures, innovative companies will create their own sub-markets within the industry; this will help to remove the current barriers to adoption for mobile business.

This chapter looks at some of the global trends behind wireless data adoption. In particular, the drivers and barriers to adoption, the telecom regulatory environment, the changes occurring within the telecommunications industry and within enterprise IT departments, the wireless Internet value chain, the wireless companies comprising the value chain, and finally some of the key applications of M-Business within the enterprise.

Many books have been written on the content of this single chapter alone. The aim here is to provide a high-level summary of some of these forces and then to move on to the strategy and implementation plans for leveraging M-Business within the enterprise for business advantage.

Global Trends

Much has been written about the global trends in the M-Business world. The United States is often cited as lagging behind the Asia-Pacific region and even further behind Europe in terms of its adoption of mobile business—sometimes cited up to two years behind. Much of this is owing to the fact that there are a number of competing wireless communications standards in the United States, versus the single standards in the rest of the world. This is actually just one of the factors that has led to Europe and the Asia-Pacific region becoming the early adopters. Other factors include cultural aspects, geographic aspects, political and regulatory aspects, pricing factors for Internet access, and the penetration rate of the wired Internet within these countries.

To understand the global trends in wireless communications and the growth of the wireless Internet, we need to start by understanding the growth of the Internet itself. Figure 2–1 shows the Internet penetration by region from a study by the ARC Group.

Figure 2–1 Internet User Penetration by Region. Source: ARC Group.

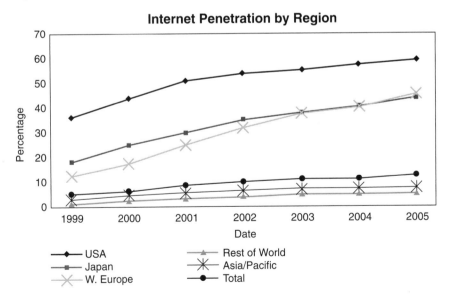

It is clear that the United States has dominated and will continue to dominate the statistics for the highest percentage penetration by region. Japan and Western Europe follow closely behind with the Asia-Pacific region and the rest of the world being further behind in penetration.

If we now turn to the penetration rates in terms of mobile data penetration (Figure 2–2), we see a different picture. The United States clearly lags behind Western Europe and Japan. Mobile data in this case includes access to data by cell phones, PDAs, and interactive pagers. At the current point in time, Western Europe is clearly the leader.

Equipped with these predictions, the questions still remain as to what services will see the most demand and how often subscribers will use the wireless data features of their devices even if they are subscribed.

Beyond looking at penetration rates by region for the Internet and for mobile data, we also need to look at the number of mobile handsets being shipped, the number of users accessing various forms of mobile data, and the number of users conducting mobile commerce.

Figure 2–2 Mobile Data User Penetration by Region. Source: ARC Group.

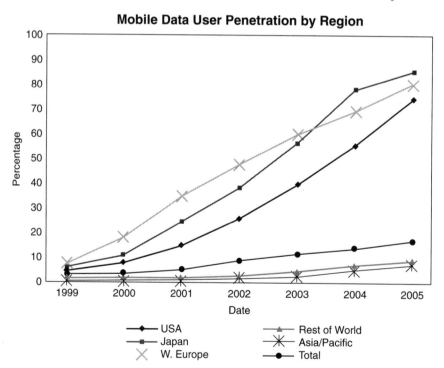

According to the research firm Jupiter, there will be 1 billion wireless Web devices in circulation by the year 2003. They also go on to say that companies must enable wireless extensions during the next 12–18 months, or risk losing customers to competitors that do.

The wireless data market has really been ignited by consumers, but it is likely that the eventual winners will be enterprises that leverage the technology within their enterprise to create substantial returns on investment. Because of this consumer-based origin of the wireless data market it is important to look, at least briefly, at some of the consumer statistics before continuing our main enterprise focus throughout the course of the book.

Table 2–1 presents more data points in terms of predictions for the number of users and revenues generated via wireless devices and M-Commerce transactions.

The analyst predictions provide some good quantitative data around the adoption of wireless data services and M-Commerce applications throughout the world. Consumers will gain access to Internet capable

Table 2–1 Predictions for Wireless Data and M-Commerce

Category	Analyst Projection
Mobile Internet Devices (Worldwide)	>> 1B mobile Internet access devices by 2003—Yankee Group
Wireless Internet Users (Worldwide)	>> Growth from 46.3M in 1999 to 1.02B in 2005—ARC Group
Enterprise Wireless Enablement	>> Enterprises will spend more than $400M by 2001 to wirelessly enable their business—Aberdeen
M-Commerce Users (Worldwide)	>> Growth from fewer than a thousand users in 1999 to 29M in 2004—IDC
M-Commerce Revenues (Worldwide)	>> $21B in revenues in 2004—IDC >> U.S. revenue generated through mobile devices by 2005: 32 billion—Merrill Lynch

devices, will then begin to subscribe and use these services, and finally will become true M-Commerce users generating M-Commerce revenues.

Searching for the Killer Application

An often-asked question within the wireless Internet community regards the killer application. Is there a killer application, and if so, what is it? The answer is that killer applications for the wireless Internet vary by culture, by country, and by individual user. In Europe, the killer application has been Short Message Service (SMS) text messaging, in Japan interactive games and pictures via the NTT DoCoMo i-mode service, in North America e-mail via 2-way interactive pagers such as the RIM BlackBerry plus WAP-based wireless data portals providing news, stocks, and weather information.

Undoubtedly, these so-called killer applications will take on different forms as the wireless networks mature, devices morph into better form factors and capabilities, and wireless carriers experiment further and build upon their lessons learned. What is certain is that the amount of content and applications available via these devices will proliferate and M-Commerce services will evolve along with the non-transactional services.

Evolution to 3G Networks

In our discussion of the global trends within the M-Business environment, one topic we will hear a lot about is that of so-called 3G or Third Generation Networks. This is a term frequently used by the wireless carriers in order to describe their next generation wireless networks for voice and data communications. In this section, I'll provide a short definition of the characteristics of 3G networks when compared to older networks such as 1G, 2G, and 2.5G (Figure 2–3). This will equip us with some of the terminology that we need to understand when discussing trends in the telecom environment and how this will affect the enterprise moving forwards.

The main advantages of the move toward 3G networks are the increased bandwidth and the worldwide standardization that 3G will bring to the global telecommunications industry. As such, increased bandwidth will enable the mainstream use of multimedia applications such as streaming audio and video and large file transfers.

Figure 2–3 Comparison of 1G, 2G, 2.5G, and 3G Networks. Source: Nokia and 3G Newsroom.

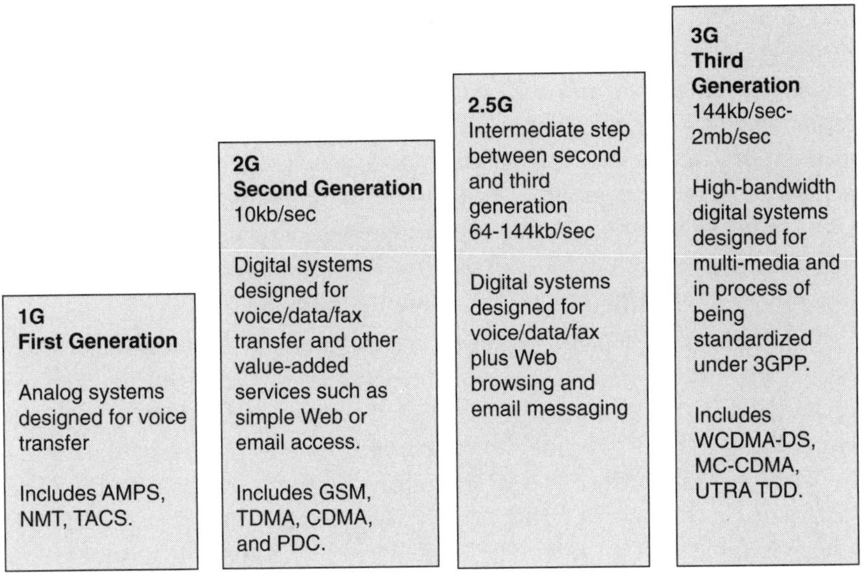

1G First Generation	2G Second Generation 10kb/sec	2.5G Intermediate step between second and third generation 64-144kb/sec	3G Third Generation 144kb/sec- 2mb/sec
Analog systems designed for voice transfer	Digital systems designed for voice/data/fax transfer and other value-added services such as simple Web or email access.	Digital systems designed for voice/data/fax plus Web browsing and email messaging	High-bandwidth digital systems designed for multi-media and in process of being standardized under 3GPP.
Includes AMPS, NMT, TACS.	Includes GSM, TDMA, CDMA, and PDC.		Includes WCDMA-DS, MC-CDMA, UTRA TDD.

Applications by Region

We'll now take a look at Europe, the Asia-Pacific, and the North American market to understand some of the wireless data applications that have obtained traction with subscribers. Since many enterprises have a business-to-consumer focus, it is useful to know what types of applications are experiencing uptake and which others are maturing.

Europe

European countries have had the advantage of a single digital mobile telecommunication standard in the Global System for Mobile communications, or GSM. GSM is a 2nd generation digital standard that accounts for over 64% of the world's wireless market. So-called 1st generation systems were the analog communications standards such as the Analog Mobile Phone System (AMPS).

GSM has international roaming capability and is supported in over 159 countries. It offers voice telephony services, including call waiting, call hold, call forwarding, and calling line identity (CLI), together with data services such as short messaging service (SMS), wireless application protocol (WAP), and general packet radio services (GPRS).

Short Message Service (SMS) has been the killer application in Europe, with over 50 billion global text messages sent within the first quarter of 2001 as reported by the GSM Association. In the UK, they report that customers generated 3.5 billion text messages in the first four months of 2001. The medium has proven popular not only for person-to-person messaging, but also as a response vehicle for television shows such as MTV that encourage audience participation. Additionally, brands such as Coca Cola and Budweiser have been leveraging the medium for targeted marketing campaigns.

The UK has also seen several M-Commerce trials and production deployments taking place. An example is the shopping service provided by the Safeway grocery chain that allows shoppers with Palm Pilot PDAs, provided by Safeway, to manage their shopping lists and submit orders to the store for picking and packing by store staff prior to customer collection. This program dates back to 1999, when Safeway offered their "Easi-Order" shopping service with Palm Pilots to 200 regular users of their "Collect & Go" home ordering service at a store near London. Safeway has since expanded the trial to more stores and customers and has plans for wireless access to the application functionality in addition to the current telephone dial-up access.

Asia Pacific

One of the biggest success stories for the wireless industry has come from the Asia Pacific region. The story and the success of the NTT DoCoMo i-mode service has been played back time and time again. NTT DoCoMo is Japan's largest mobile operator and has 24 million customers using the i-mode service. The i-mode service employs packet data transmission. Communications fees are charged by the amount of data transmitted/received rather than the amount of airtime.

Some of the services available include mobile banking, travel reservations, restaurant/town information, message services for news, I-mode compatible Web sites, e-mail, entertainment sites such as Disney and Universal Studios, and downloadable ring tones. DoCoMo provides certain content for free and provides premium content and applications for a monthly fee that ranges from 100 to 300 yen per month per offering.

One of the most interesting things about the I-mode service has been the speed with which consumers have adopted the service. The service started on February 22nd 1999, hit the 5M subscriber mark around the 1st year of service, and the 20M subscriber mark around the 2nd year of service. The adoption rate and revenues generated have been the envy of wireless carriers around the world.

The introductory phase of the company's "FOMA" 3G rollout was heavily over-subscribed with applications for nearly 150,000 mobile phones with 4,500 actually given out. Of these 4,500 mobile phones in trial, 1,200 were "visual" phones equipped with a video screen. FOMA is NTT DoMoCo's name used in Japan for their W-CDMA services and stands for "Freedom Of Mobile multimedia Access."

As 3G trials and rollouts move forward with the Asia Pacific region and within Europe, carriers within the United States are able to gain an early view into the adoption patterns for these types of services and adjust their strategies accordingly.

North America

In North America, we have witnessed the popularity of the Sprint PCS wireless Web together with similar offerings from AT&T Wireless, Cingular Wireless, and Verizon Wireless among others. Sprint PCS passed the one million subscriber mark for wireless Web customers within the first year of its service.

In addition to access to the Internet via WAP-enabled cell phones, which is still a maturing application in the United States, one of the big trends in the U.S. has been the use of RIM wireless handhelds for receiving and sending corporate e-mail. The RIM 950 and 957 wireless handhelds manufactured by Research In Motion (RIM) provide an always-on service for wireless e-mail using the DataTAC and Mobitex wireless networks. Network operators for these services include Motient Corporation and Cingular Interactive in the United States and Bell Mobility and Rogers AT&T Wireless in Canada. Revenues for the operators of these services are attractive with monthly charges of $30 for 100,000 character service fairly typical.

Task-To-Device Affinity—
WAP Phone and RIM Pager Comparison

A personal anecdote may be useful in explaining what Forrester Research has termed the task-to-device affinity for wireless devices. This may help to explain why certain applications have been so successful with consumers and business users.

In addition to regularly using a laptop and PDA, I have a RIM pager and WAP-enabled cell phone. When comparing the usage levels of wireless Web against interactive messaging, I personally find myself spending more time with my RIM pager than with the data features of my WAP phone. One of the reasons, I believe, is due to the ease of use factor, or as Forrester terms it, the task-to-device affinity.

The RIMs' keyboard makes composition of e-mail messages very easy and much simpler than the equivalent process on a WAP phone. I tried sending e-mail over my WAP phone when I first obtained the phone and wanted to experiment. The process was so difficult that it took several minutes to compose a simple e-mail message and dispatch it. Conversely, I have found my WAP phone most useful for data access. Looking up stock quotes and news items and any tasks that do not require heavy text input. Even reading e-mail messages is acceptable on the WAP phone—the only limitation is really the data entry portion at present.

The task-to-device affinity issue is certainly a moving target. As cell phones and PDAs evolve into smartphones that combine the best of both worlds, the devices become more useable for a variety of functions including voice, e-mail, and Internet access. Today, WAP phones are good for data access, but not for all forms of data entry. Conversely, pagers are good for the single function of sending and receiving e-mail. The task-to-device affinity is an important topic especially for the enterprise since the ability to consolidate from three or four devices down to two or three can yield substantial cost savings in support costs.

Drivers and Barriers to Adoption

Drivers for Adoption

The drivers for adoption of mobile business within the enterprise and with consumers are numerous (Figure 2–4). They include the following: the increasing mobility of today's workforce; the convergence of telecommunications and software industries; the increasing need for information and transactions anytime and anywhere; the new breed of wireless handsets coming on the market; the revenue opportunities created via location-based services and M-Commerce; the productivity improvements to be gained via wireless extensions to enterprise applications and processes; the improvements in bandwidth brought about by the migration from 2G to 2.5G and 3G networks; and the adoption of wireless standards such as WAP and Bluetooth, together with the cultural and regulatory drivers in various countries. If we distill these drivers into their primary forces, we see the forces of industry convergence, improvements in wireless technology and standards, together with cultural and regulatory effects as driving global adoption of mobile business.

Figure 2–4 Drivers for Adoption of Mobile Data.

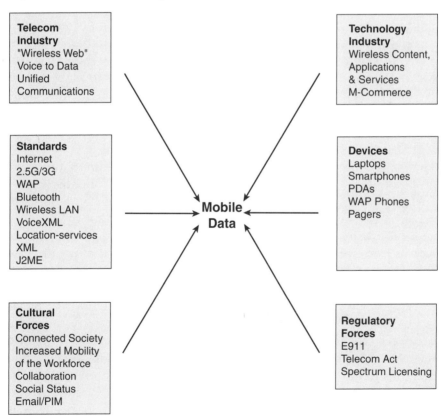

Barriers to Adoption

Despite the strong forces that are driving adoption of mobile business, it also faces considerable barriers. There are two main types of barriers to adoption: business barriers and technology barriers. The primary driver for adoption of any new technology needs to be the business case. But even with a business case developed, if the technical obstacles are too high, deployment will be troublesome, if not impossible.

As technical obstacles diminish, they can actually help the business case by expanding the realm of possibilities. For example, accurate location determination techniques can create opportunities for location-based advertising.

Business management needs to be aware of both the drivers and the barriers to adoption of mobile business so that informed decisions can be made. Acting too early or too late can have significant consequences. An entry into mobile business that is premature or incorrectly targeted can distract scarce resources within the enterprise · without achieving significant results. An entry that is too late can be even more dire and lead to lost revenues, lost productivity, lost competitive advantage, and even lost customers.

Business Barriers

The major business challenge for mobile business is simply the business case. On a macro scale such as the creation of an entirely new business, the following standard questions may apply: Can a business make money by using this model? What is the nature of the product or service being offered? Who are the customers and how will they benefit from this product or service? What is the point of pain that is being removed? What is the size of the market and the differentiation from the competition? What is the pricing strategy and how will the service be delivered? What channels will be used to promote the product or service? What should the branding strategy be? Are end users ready for this service? These are all fairly classic questions for any business. They apply equally to the M-Business arena because the market still has to be created and moved from early adopter status to the mainstream. Often, in addition to a compelling value proposition around M-Business, one still needs to educate and influence consumer and enterprise behavior in order to drive adoption. The new way of doing business needs to be compelling enough and simple enough in order to change user behavior.

On a smaller scale, such as a new business initiative within an enterprise, the following questions may apply: Will end-users accept the technology and process change? Will it provide enhanced customer service or improved employee productivity? How will this be measured? How will end users transition from prior processes into this new process? What is the return on investment? What is the learning curve for end-users? What training is required? What support services are required? What service level agreements need to be in place? How critical is this new application to the business?

The case studies in the industry examples chapter of this book will serve to illustrate how some of the enterprise early adopters have answered these questions and have achieved true business benefit and return on investment.

Technology Barriers

The technical barriers to adoption of wireless technologies are numerous. They include diverse standards for applications and networks, spotty coverage, low bandwidth, perceived lack of security, diversity of devices, slow response times, primitive user interfaces, and numerous other factors.

In a December 2000 survey of 101 IT and business managers, Internet Week found the following distribution of wireless Internet concerns (see Table 2–2).

Table 2–2 Wireless Internet IT Concerns. Source: Internet Week.

Rank	Feature	Percentage
1	Security	77%
2	Lack of Reliable Standards	69%
3	Lack of Web or Enterprise Integration Products	61%
4	Inadequate Bandwidth	54%
5	High Costs of Technology	49%
6	Quality of Technology	44%

The concerns are similar to those of the wired Internet about four years ago. Typical concerns back then included the primitive graphical user interface of the Web browser versus the richer user interface of client/server applications, the lack of security, and the low bandwidth. Enterprises were not convinced that the Internet technologies were robust enough for their critical applications. In fact, I remember many meetings with enterprise clients as a consultant where the stakeholders questioned the need for applications such as extranets and quite rightfully asked about the return on investment. Since ROI calculations had not been extensively developed in those early days, we tended to talk about the soft benefits of enhanced customer satisfaction and improved communications.

Moving back to the present day, as the industry continues to evolve, innovative technology companies and wireless carriers are providing solutions to these technology obstacles—thus helping to drive adoption. What is clear is that the enterprise cannot afford to wait. Mobile business strategies should be crafted today in order to target quick wins and to drive the process change within the enterprise toward mobile business.

Even with a lack of reliable standards, inadequate bandwidth, incomplete coverage, and a wealth of devices and software on the market, it is possible to design and implement highly effective applications within the enterprise that provide a good return on investment. Applications can be implemented that support multiple devices, multiple carrier networks, and can handle incomplete coverage by offering online and offline capabilities. Typically, during offline usage where the carrier network cannot be accessed, the applications use the onboard database of the device and store data for later synchronization when the wireless network becomes available or when a standard dial-up connection or cradle connection is available.

Regulatory Environment

In addition to the major global carriers, handset manufacturers and software companies creating the market for mobile business for the enterprise, the global regulatory environment is also helping to chart its course. Regulations such as the Telecommunications Act of 1996 and the Enhanced 911 (E911) mandate from the Federal Communications Commission (FCC) within the United States have helped to bring about major change in the telecommunications industry.

The FCC was established by the Communications Act of 1934 as an independent United States government agency directly responsible to Congress. The FCC is responsible for establishing policies to govern interstate and international communications by television, radio, wire, satellite, and cable.

Enhanced 911

In the United States, the Federal Communications Commission's E911 mandate made automatic location identification a requirement for the wireless carriers to implement within their networks. The following is an extract from the FCC Web site:

> *"In a series of orders since 1996, the Federal Communications Commission (FCC) has taken action to improve the quality and reliability of 911 emergency services for wireless phone users, by adopting rules to govern the availability of basic 911 services and the implementation of enhanced 911 (E911) for wireless services."*

The basic 911 rules required wireless carriers to transmit all 911 calls to a Public Safety Answering Point (PSAP) without regard to validation procedures intended to identify and intercept calls from non-subscribers. Phase I of the enhanced 911 (E911) rules, required carriers to provide to the PSAP the telephone number of the originator of a 911 call and the location of the cell site or base station receiving a 911 call. Phase II of the E911 implementation required wireless carriers to provide Automatic Location Identification (ALI) beginning on October 1, 2001 in order to provide emergency services with greater accuracy for call origination. The ALI accuracy requirements were as follows:

>> For handset-based solutions: 50 meters for 67% of calls, 150 meters for 95 percent of calls

>> For network-based solutions: 100 meters for 67% of calls, 300 meters for 95 percent of calls

This Government mandate has helped add fuel to the location-based services industry as a sub-set of the M-Business market. According to Strategy Analytics, the market for location-based services will reach $6.5 billion in the United States and $9 billion in Europe by 2005. Some of the potential applications of located-based services include tracking services for locating and tracking people and assets, and location-based advertising. The business models and potential applications for location-based services will be explored in further detail later in the book.

Telecommunications Act of 1996

The Telecommunications Act of 1996 was the first major overhaul of telecommunications law in almost 62 years within the United States. The goal of the law was to let anyone enter any communications business—to let any communications business compete in any market against any other.

> *"To promote competition and reduce regulation in order to secure lower prices and higher quality services for American telecommunications consumers and encourage the rapid deployment of new telecommunications technologies."—Telecommunications Act of 1996*

The main thrust of the law was to force the Bell Operating Companies to open up their local loops to competitors in exchange for providing them the ability to enter the long distance market. The Telecom Act has done a lot to create a competitive environment in the telecommunications industry and has resulted in a large increase in financial investments in the telecom industry.

Spectrum

The allocation of spectrum, the various frequencies for radio transmission, is also subject to regulation. Radio spectrum is the part of the natural spectrum of electromagnetic radiation lying between the frequency limits of 9 kilohertz and 300 gigahertz.

The International Telecommunications Union (ITU) in Geneva is responsible for worldwide coordination of both wired and wireless telecommunications activities. Frequency planning is conducted by the ITU Radiocommunication sector (ITU-R), which has divided the world into three broad regions.

In the United States, responsibility for radio spectrum is divided between the FCC and the National Telecommunications and Information Administration (NTIA). The FCC administers spectrum for non-Federal government use and the NTIA, which is an operating unit of the Department of Commerce, administers spectrum for Federal government use.

3G Licensing

One of the biggest costs, in addition to physical infrastructure building, for the wireless carriers in moving toward 3G networks has been
the bidding on spectrum auctions. According to Nokia, over 80 3G
licenses were granted in 2000 and several hundred more will be
granted over the next few years.

The U.K. government auction, which ended on 28th April 2000
after seven weeks of bidding, raised nearly 22 billion pounds from five
operators—TIW, One2One, Orange, Vodafone, and BT Cellnet. The
German auction, which ended on 5th September 2000 after fourteen
days of bidding, raised $37 billion from six operators—T-Mobil,
Mannesmann, E-Plus-Hutchison, Viag Interkom, MobilCom, and
Group 3G. According to 3G Newsroom (http://www.3gnewsroom.
com), the five most expensive auctions were Germany, Britain, USA,
Italy, and South Korea with a total of over $100 billion spent within
these five countries alone.

Other countries such as Norway and Finland have adopted a
beauty content approach and have given away spectrum licenses for
free or for reduced prices when compared to auction pricing. Due to
the high fees paid out by the wireless carriers in buying these 3G
licenses, companies such as BT and Vodafone have actually started to
talk about teaming in order to share the costs of the 3G buildout in
terms of network infrastructure such as mobile masts.

With this much money invested in 3G networks, it will be interesting to see how quickly the wireless carriers transform their business
models, products and services, and target both consumers and the
enterprise in order to attempt to recoup these massive expenditures.

Telecom Environment

One of the major trends in the telecom industry is that the wireline
and wireless carriers are seeking to leverage data services as a new
value-added service offering, a new revenue stream, and a differentiator from the competition. With data services, these companies have
the potential to increase average revenue per user (ARPU) and reduce
customer churn. With voice service becoming increasingly competitive
and commoditized, these companies are moving into data services
with the same zeal as enterprises that have adopted E-Business over

the last several years. In effect, they are transitioning themselves into a totally new class of service provider—often termed the next generation communications service provider or CSP.

The following figure from the ARC Group shows how large the problem of declining voice revenues is becoming. The next year or two will be critical for the communications companies to transition their business model and their products and services in order to capture the market for data services before it is lost to the competition.

Value-added data services can help communications companies not only increase ARPU but also to reduce customer churn and to increase customer loyalty. Consumers, small businesses, and enterprise customers are far more likely to stay with a given service provider if they depend upon them for not just their "pipe" (the connection) but also productivity increasing communications services. Such services include unified messaging and e-mail, content services that are personalized to their needs, and applications such as personal

Figure 2–5   Predictions for Increase in Value-Added Data Services in Order to Raise Carrier ARPU. Source: ARC Group.

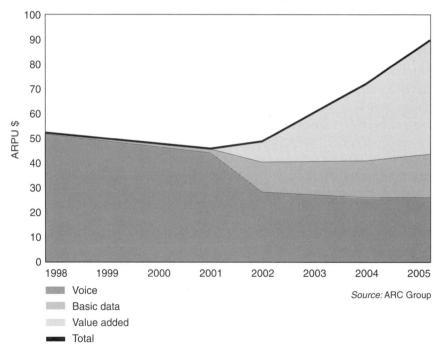

information management, Internet and intranet access, time and expense reporting, dispatch/scheduling, equipment monitoring, sales force automation, and field force automation.

Today's communications companies could end up becoming the M-Business providers of tomorrow. Owning the network, they are in a powerful position to become the ultimate owner of the customer— financial services providers, retailers, content companies, and software companies need to pay close attention to the strategies and movements of some of the leaders in this industry.

In fact, we appear to be moving into an era of vertical integration. Wireless companies are attempting to become full service providers within the industry. They are beginning to do this by providing the network, the applications, and the integration services for their enterprise customers. An example that comes to mind is the merger of Motient Corporation, a provider of two-way mobile and Internet communications services, with Rare Medium, an Internet professional services firm. This vertical integration model is reminiscent of the early decades of the computer industry, wherein companies such as IBM, HP, and Digital provided complete solutions from hardware to software.

Wireless Data Services

One of the first data services, after two-way interactive paging, provided by the wireless carriers was the well-known "wireless Web." This service provided access to Internet sites over a cell phone equipped with a micro-browser via the Wireless Application Protocol or WAP. The WAP standard defines a set of technical specifications for delivering Internet communications and advanced telephony services on digital mobile phones, pagers, personal digital assistants, and other wireless terminals. The initial work on WAP started back in June 1997; this was done by a consortium that included Ericsson, Motorola, Nokia, and Unwired Planet. (Since that time Unwired Planet changed its name to Phone.com and later merged with Software.com in order to form Openwave.)

Thus, the first generation of wireless Internet data services provided by the carriers was focused on consumers and provided access to Internet Web sites via the Wireless Application Protocol. Carriers formed partnerships with multiple content providers in order to build closed-content portals often known as a "walled garden." Users were confined to the content providers listed within the portal and had little or no way of getting out to any other sites on the Internet. Given the "walled garden" situation, it has made a lot of sense for content providers to form relationships with the wireless carriers in order to get placement on their WAP portals and better visibility for their wireless services.

The story of how wireless carriers came to focus on horizontal wireless data applications for enterprise customers is an interesting one. According to Paul Reddick, VP of Business Development, Sprint PCS, the company was already looking at enterprise applications for wireless data as early as 1998. They initially talked to enterprise customers to understand the killer applications for various vertical markets such as the insurance and real estate industries. Their discussions with customers revealed that, although vertical market applications did not warrant heavy investment at that time, there appeared to be a universal need for more horizontal applications on the mobile phone such as e-mail, and personal information management (PIM) such as contacts, calendars, schedules, as well as customer relationship management, corporate directories, and sales automation. Sprint PCS adjusted their strategy and delivered these horizontal applications with a set of industry partners as part of a suite of products known as the Wireless Web for Business in September 2000. They are now moving into the vertical applications space as described in the next section on wireless application service providers. According to Reddick, part of mission of the Sprint PCS Clear Wireless Workplace is to make people more productive. The natural extension to this is to allow their customers to access information in addition to people. This is true whether one is dealing with consumers or with enterprise customers. Thus, the migration from voice to data services has been a natural evolution, as well as a fundamental part of their mission statement.

M-Commerce Services

The next logical step after wireless data services such as wireless Internet access is for the wireless carriers to facilitate M-Commerce transactions over their networks. Many pilots and trials have been adopted worldwide, with some carriers such as NTT DoCoMo already having production implementations. We have seen this earlier in the section on global trends. Examples of some of the trials that have occurred include the NetCom trial with the M-Commerce software company MoreMagic, the AT&T Wireless trial with QPass, and the DirectBill service offered by Cingular Wireless.

The trial conducted by Norwegian mobile operator NetCom used the MoreMagic payment transaction software to pilot four M-Commerce service offerings: a popular Norwegian daily soap, a pizza delivery service, an online newspaper archive, and a location-based service. The MoreMagic transaction platform is described in Chapter 8.

Outside of the wireless carriers, other players are also engaging in pilots. Palm has been testing M-Commerce payments using the Palm PDA as the holder of digital wallet information; such information can be beamed over the IR port to merchants with Palm-compatible terminals for payment.

Since M-Commerce is still in its infancy, business models have not yet stabilized. It remains to be seen who the eventual winners will be. The main players are the wireless carriers, the portals and content aggregators, the financial service institutions, and the merchants themselves. How the value extracted from M-Commerce transactions will be shared between these players is still to be determined. What is likely is that a significant portion of the percentage of the revenues will shift from the wireless carriers toward the content and application service providers.

Who will own the M-Commerce consumer is also an open question. The carriers, as we suggested earlier, may be one possibility. They control and operate the network and the portal interface presented to the user. But the financial services institutions, who can be carrier network agnostic, may also be able to extend their customer relationship from the credit card world to the M-Commerce world, that is with digital wallets. The digital wallet is an important item to own from a provider standpoint, because it can contain customer payment choices and shipping addresses, as well as customer preferences and a link to the customer's transaction history and buying habits.

One aspect of M-Commerce transactions that the wireless carriers have seemed reluctant to own is the billing for third-party products and services. The issues around billing relate to the legality of billing for non-telecom-related changes, the issue of collecting payment, and the issue of customer care and dispute resolution.

The winning strategy for the wireless carriers may well be to out-source the billing and collection aspects around M-Commerce trans-actions, but to own the customer profile, preferences, and the digital wallet. This builds in switching costs for the consumer, owing to the time required to activate a digital wallet, and the level of personaliza-tion and ease-of-use that it provides. Yet this still frees the wireless carrier from the burden of handling or providing all the costs associ-ated with post-transaction customer care and billing.

On the other hand, merchants may become less willing to share rev-enues if all the carrier provides is the channel to the customer. As closed wireless carrier portals give way to open portals with free access to any Internet URL, the carriers may find themselves dis-intermediated from M-Commerce transactions with merchants who already have a strong brand name and customer loyalty. To stay in the loop, they need to pro-vide more user-friendly, efficient, and secure M-Commerce mechanisms than the merchants provide by themselves. This may include digital wal-let services, one-click transactions, ease of navigation, security, and con-text-relevant services that enhance the value proposition for both the merchant and the consumer. This model of value-added services in order to stay a key component of the value chain is similar to the strategies of distributors in the supply chain of the business-to-business electronic commerce world.

The action item for the enterprise contemplating their M-Commerce strategy is to continue to observe the various business models and rev-enue sharing arrangements that are occurring worldwide. Moreover, enterprises need to be prepared for when this channel becomes signifi-cant. Eventually the M-Business channel will be just as important as the wired Internet channel to customers that you have today. Today, there are few consumers making M-Commerce transactions, so the incentives for the enterprise to invest in and roll out M-Commerce solutions are reduced. However, it is prudent for the enterprise to plan an overall M-Business strategy that considers customers, employees, suppliers, and business partners. The evolution and adoption of M-Commerce within those user constituencies must be a consideration in an enterprise's future endeavors. Be prepared to migrate from wireless communications and content to wireless commerce as your customers begin their adoption. Providing simple non-transactional M-Business services to customers today can also help to pave the way to transactional M-Commerce inter-actions with your customers in the future.

Wireless Data Example

AT&T Wireless

http:www.att.com

As an example of a wireless Internet offering provided by a major carrier, we'll take a look at the Digital PocketNet Service offered by AT&T Wireless. In addition to access to WAP-enabled Web sites, this service provides e-mail, address book, calendar, alerts, and to-do functionality for the cellular phone. The e-mail account has the format username@mobile.att.net and users can customize the settings of their wireless data services either directly on the cellular phone or via the AT&T Web site at http://www.att.com/mypocketnet. Updates to the preferences made on either the cell phone directly or via the Web site are reflected immediately in the service. For example, the Web site can be used to enter favorite links to Web sites or favorite phone numbers. The personal Web site provided by AT&T Digital PocketNet Service is powered by InfoSpace. Additionally, FoneSync software from Openwave is used to provide synchronization capabilities between the PocketNet Service and a user's Personal Information Management (PIM) software, i.e., Microsoft Outlook, Lotus Notes, Lotus Organizer, Symantec ACT!, and Goldmine.

Figure 2–6 shows some of the menu options and partner content available from the "Web Sites" section of the PocketNet service.

Figure 2–6   Sample of Content Providers on the AT&T Digital PocketNet Service.

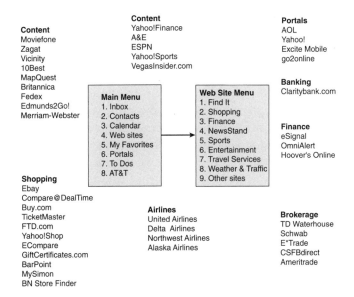

Wireless Application Service Provider Platforms

The next generation of wireless Internet data services provided by the wireless carriers has started to focus on the enterprise. One of the major trends has been for the wireless carriers to provide wireless application service provider (or WASP) services to enterprise customers. This gives an enterprise wireless access to a variety of applications and content. In 2001, almost every major wireless carrier in the United States announced plans to provide these WASP services to the enterprise by teaming with systems integrators and software companies.

It should be noted that prior to the wireless carriers entering the market, many wireless software vendors had been providing their technology as a hosted service to the enterprise; this had initially created the wireless application service provider market. Such companies included 2Roam, Aether Systems, Air2Web, and JP Mobile.

As examples of wireless carriers entering the WASP market, Sprint PCS announced a relationship with Compuware to deliver applications such as wireless meter reading, real-time tracking for the transportation industry, mobile phone applications, enterprise resource planning and customer relationship management tools. AT&T Wireless announced a relationship with Accenture to wirelessly enable applications such as corporate e-mail, customer information management, sales force automation, and inventory management. Cingular Wireless announced a relationship with Siebel Systems to offer Siebel eBusiness Applications across their network and a Corporate E-mail PLUS service to give enterprise employees wireless access to their Microsoft Exchange or Lotus Notes-based corporate e-mail systems.

Many business challenges still exist for the carrier versions of these wireless application service providers to become a hit with the enterprise. Their go-to-market strategy has to make sense for the enterprise buyer from both a business and technology standpoint. The business buyer needs to feel assured that the wireless application service provider understands their business and has a solution that can add real value. Business buyers may feel a little uncomfortable buying services from a wireless carrier who is unfamiliar with their industry and business processes. IT buyers need to feel assured that the offering provides all the required levels of security, performance, reliability, and scalability, in addition to contract flexibility and the breadth of the service and support offerings. When faced with selecting a single wireless carrier for wireless access to enterprise applications, the issue of coverage will also be very important.

Finally, the wireless application service provider will need to have smoothed out all the internal partnerships that need to occur to provide an end-to-end service offering for the enterprise. This will typically include relationships with systems integrators and a number of software vendors. One of the fundamental challenges is that wireless carriers do not typically understand enterprise needs in terms of application solutions. If wireless carriers can successfully cross over this hurdle, they will be well on the way to becoming valuable service providers for the enterprise and will have successfully broadened their data services from a consumer focus to an enterprise focus.

IT Environment

As we have discussed in Chapter 1, the IT department is also changing rapidly. Why are changes to the IT department important to an executive considering his or her business strategy for mobile customers, partners, or employees? By understanding the dynamics occurring within these departments, the business executive can have a better set of expectations around what can be delivered, how quickly, and how the progress of an initiative can be tracked and measured.

IT was under an increasing amount of scrutiny from the business side during the slowdown in the economy in mid- to late-2000. At the same time, IT departments were becoming increasingly complex in terms of application portfolio, technical architectures, hardware and software configurations, departmental skill sets, best practices, methodologies, standards and guidelines, and development lifecycles. The ratio of boxes (servers) to IT staff has been increasing and many enterprises and service providers have evolved to the "thousand server environment"—where literally thousands of computer servers are required to run the business.

Why was IT never scrutinized over the past couple of years? Mostly because the economy had been so strong that IT spending was questioned, but not scrutinized. The downturn in the economy in late 2000 and early 2001 caused IT business management practices to be improved and IT expenditures to be more justified. The total cost of computing and return on investment is being more accurately tracked, reported, and articulated to the business.

It is important to note that if we look at the changes within the IT department on a longer time scale (that is, in terms of decades), we find that this scrutiny in terms of costs and justification for projects is actually a return to normalcy. In fact, the anomaly was the couple of years during the Internet and E-Business era of the mid to late 90s, where spending was increased and the race was to capture the market and stay ahead of the competition.

These forces of increasing scrutiny from the business side, together with increasing complexity and the evolution of IT from a cost center to a strategic partner, have meant that mobile business initiatives and other initiatives are now becoming more carefully designed, developed, and deployed. Moreover, IT is now equipped with additional, if perhaps rudimentary, tools for the calculation of total cost of computing and the return on investment for IT initiatives.

IT has also shifted some of its focus from revenue generation activities and applications toward applications that can help to improve productivity and reduce costs for the enterprise. It is here that M-Business initiatives can have their first impact upon the enterprise. Later, as the technologies mature and consumer acceptance and penetration increases, M-Business initiatives can play an equal role as a source of revenue generation.

When looking at the relationship between IT and the business side, it is worth noting the top ten problems that customers have reported with their IT organizations. The following top ten list is taken from a report by the Giga Information Group:

>> The value of IT is not understood

>> Projects are not executed properly

>> Project selection is wrong

>> IT means business prevention

>> IT cannot maintain a robust infrastructure

>> IT leadership does not contain costs

>> IT is not business savvy

>> IT does not establish and maintain an effective dialog with the rest of the business

>> Results need to be reported better

>> IT people are not customer oriented

The takeaway for the business executive considering the potential of mobile business for his or her employees, customers, suppliers, and business partners, is to ensure a strong dialog and relationship with IT. The dialog should be built around the goal of creating enterprise value through the leverage of IT resources.

Wireless Internet Value Chain

When looking at the current state of M-Business, it often helps to view the entire value chain in order to see how companies are positioning themselves within their markets. The wireless Internet value chain appears to be a lot more complex than the E-Business value chains that have preceded it. There are more moving parts and more players involved. As we migrate from 2G to 3G wireless networks, the value chain complexity increases even further, as more application and content providers take advantage of the higher capabilities of the networks.

Another part of the complexity of the value chain is owing to the number of standards and technologies involved. Not only do we have Internet hardware and software vendors to contend with, as was the case in the E-Business value chain. Now we are dealing with device manufacturers for PDAs and cell phones, carriers for wireless networks, various standards such as WAP and WML (Wireless Markup Language), and a host of other complexities—including coverage, security, bandwidth, provisioning, billing, and both voice and data applications.

As mobile business extends the enterprise to any-location and any-time access, we now start to see how established industries can play a more integral part in the value chain. For example:

>> A commercial real estate property owner can gain revenue by placing wireless technologies that boost cellular carriers' signals inside a building. Inner Wireless is an example of a company that provides passive cell tower technologies to the commercial real estate industry. The technology has been in field tests. It has proven successful not only within office spaces, but also within the parking garages and elevators of said properties. This technology solves a large problem cellular signals just one foot inside a building are often 100 times less than those just one foot outside the building with a corresponding effect on the service quality provided to the end user.

>> Airline carriers can provide wireless LAN capabilities in their lounges and at the gate. Using the MobileStar service in several airports in the United States, American Airlines has rolled out wireless LAN service to gates and Admirals Club lounges. This is being done in New York (JFK), Newark, San Francisco, Chicago (O'Hare), and Baltimore/Washington.

>> Automotive manufacturers can build in telematics services as a value-added technology within their luxury cars. Telematics services from companies such as ATX Technologies provide drivers with emergency services (such as automatic collision notification, emergency response, and roadside assistance). They also provide navigation and information services, such as routing assistance, traffic, weather, news, financial information, and sports information—all of which can be defined and customized from a personalized Internet page.

>> Coffee shops can provide wireless LAN capabilities. Starbucks, along with its partners Compaq and MobileStar, is starting to offer wireless LAN capabilities to customers. Customers provide their own laptops or PDAs, but are able to use the Starbucks wireless LAN access point. Pricing models for these kinds of services appear to be evolving, but can either be an hourly rate or a flat rate per month.

Value Chain for Enterprise Wireless Data

Figure 2–7 shows a typical M-Business value chain from the network operator to the enterprise customer.

In this sample value chain, we can imagine the case of an enterprise giving wireless access to e-mail, Internet, and intranet content to their employees. In this case, we can truly see how extensive any given value chain can become. The network operator may be a wireless carrier such as Sprint PCS. The network infrastructure provider

Figure 2–7   M-Business Value Chain for the Enterprise.

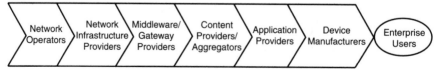

may be a networking equipment provider such as CISCO. The middleware/gateway provider may be a wireless middleware vendor such as Brience. The content aggregator may be a wireless vendor such as InfoSpace. The application provider may be a wireless application service provider such as JP Mobile. Finally, the device manufacturer may be a handset manufacturer such as Ericsson.

In this example, the enterprise customer might access a WAP-enabled Internet site via InfoSpace, his or her e-mail on Microsoft Exchange (enabled for wireless access by JP Mobile), and an enterprise application on the intranet via wireless middleware from Brience.

The major components of the value chain include the networking components (both the networks themselves and the operators of those networks), the communications software components (including infrastructure software such as wireless middleware and gateways), content and application services, and the end user devices.

If we add the physical location component to the equation, that is to say the physical point of access to the applications and services, the value chain changes. This category could be included within the application provider section of the value chain or perhaps within a new category labeled access provider.

Players within this application provider or access provider category could be the physical access providers, such as airlines and hotels, together with the located-based services companies who provide the location information for end users, target destinations, or assets that need to be tracked.

Value Chain for Consumer Wireless Data

Figure 2–8 shows a simplified mobile business value chain for the consumer.

The main difference between the consumer-focused value chain for M-Business when compared to the enterprise-focused value chain is that the consumer value chain focuses more on external retailers and

Figure 2–8 M-Business Value Chain for the Consumer.

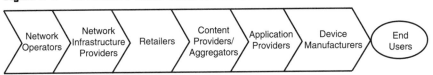

content providers. It has less to do with the internal wireless middleware and enterprise applications that characterize the enterprise space.

The value chain becomes more complex as network operators work with any array of retailers, financial institutions, content providers, and advertisers to assemble their wireless data portals for the consumer population.

Key Wireless Companies

Some of the key players who are shaping the mobile economy include the device manufacturers, equipment manufacturers, wireless carriers, wireless service providers, telematics providers, and wireless software companies. Table 2–3 shows a sample listing of some of these players. Since the market is moving so quickly and smaller players often merge, are acquired, or go out of business, this table simply shows some of the current players in the space at the time of this writing. It is certainly not to be considered an exhaustive list.

Publicly-Traded Wireless Vendors

Companies in the wireless sector have enjoyed the same roller-coaster ride in terms of their stock price as the rest of the technology industry. Some of the major players are worth studying. By doing so, we may better understand the market dynamics and the interdependencies that the various players have on one another in the value chain.

A downturn in economic outlook for the wireless carriers has a ripple effect for the telecom equipment manufacturers and the handset manufacturers. This filters down along all aspects of the value chain. The drivers of the market can be considered the telecommunications providers—both wireline and wireless. The telecom equipment manufacturers follow after them, with the software companies focused on carrier-specific wireless applications and the wireless application service providers close behind.

M-Business magazine's M-Business 50 Stock Index presents a good sampling of the major players within the mobile economy. It tracks 50 public companies, ranging from the global telecom giants to mobile startups. Table 2–4 presents an alphabetical listing of these players together with their respective stock symbols.

Table 2–3 Key Wireless Companies in the Wireless Internet Value Chain

Device Manufacturers	Wireless Middleware/ Gateways	Wireless Carriers
>> Compaq	**Gateways**	**North America**
>> Ericsson	>> Ericsson	>> AT&T Wireless
>> Handspring	>> IBM	>> Cingular Wireless
>> Kyocera	>> Microsoft	>> Nextel
>> Mitsubishi	>> Nokia	>> Qwest Wireless
>> Motorola	>> Openwave	>> Sprint PCS
>> NEC	**Middleware**	>> Verizon
>> Neopoint	>> @Hand	>> VoiceStream
>> Nokia	>> 2Roam	**Japan**
>> Palm	>> 724 Solutions	>> KDDI
>> RIM	>> Aether	>> NTT DoCoMo
>> Sony	>> Brience	**Europe**
>> Symbol	>> Cyneta Networks	>> BT Genie
	>> Everypath	>> NetCom
	>> IBM	>> Orange
	>> iConverse	>> Telia
	>> JP Mobile	>> TIM
	>> InfoWave	>> T-Mobil
	>> Microsoft	>> Virgin Mobile
	>> Oracle	**Asia/Pacific**
		>> KT FreeTel
		>> LG Telecom
		>> KTM.com
		>> China Mobile
		>> China Unicom

(continued)

Table 2–3 *(continued)*

Wireless Applications	Wireless Applications	Wireless Applications
CRM	**M-Commerce**	**Security**
>> E.piphany	>> 724 Solutions	>> Baltimore
>> eWare	>> Aether	>> Diversinet
>> PeopleSoft	>> Brokat	>> Entrust
>> Siebel	>> More Magic	>> VeriSign
FFA	>> Qpass	**Telematics**
>> @Hand	>> Trintech	>> ATX Technologies
>> Antenna Software	**Portals**	>> OnStar
>> FieldCentrix	>> AvantGo	**UM/UC**
>> Siebel	>> Handango	>> Openwave
SFA	>> InfoSpace	>> Webley
>> Pivotal	>> Openwave	**Voice**
>> SalesForce.com	>> Oracle	>> Informio
>> Siebel	>> Palm.net	>> Nuance
Alerts/Messaging/ Monitoring	>> Sirenic	>> SpeechWorks
>> CellExchange	**Location-Based**	
>> Envoy Worldwide	>> Cell-Loc	
>> Notifact	>> Openwave	
	>> SignalSoft	

Operating System & Browser Manufacturers	Wireless Equipment Manufacturers	Wireless Application Service Providers
	>> American Tower	
>> Microsoft	>> Cisco	>> 2Roam
>> Openwave	>> InnerWireless	>> Aether
>> Palm	>> Lucent	>> Air2Web
>> Symbian	>> NEC	>> Everypath
	>> Nortel Networks	>> JP Mobile
	>> Red-M	>> Seven
	>> Siemens	

Table 2–4 M-Business 50 Stock Index. Source: M-Business Magazine

Company Name	Symbol
Aether Systems	AETH
Alcatel	ALA
AT&T Wireless Group	AWE
At Road	ARDI
AvantGo	AVGO
BellSouth	BLS
Boston Communications Group	BCGI
British Telecommunications	BTY
CellPoint	CLPT
Certicom	CERT
Cisco Systems	CSCO
Comverse Technology	CMVT
Deutsche Telekom	DT
Ericsson	ERICY
Extended Systems	XTND
GoAmerica	GOAM
Handspring	HAND
Interdigital Communications	IDCC
i3 Mobile	IIIM
InfoSpace	INSP
Leap Wireless International	LWIN
Motient	MTNT
Motorola	MOT
Nextel Communications	NXTL

(continued)

Table 2–4 *(continued)*

Company Name	Symbol
Nokia	NOK
Nortel Networks	NT
Novatel Wireless	NVTL
NTT DoCoMo	NTDMY
OmniSky	OMNY
Openwave Systems	OPWV
Orange	ORNGF
Palm	PALM
Pumatech	PUMA
Qualcomm	QCOM
Research in Motion	RIMM
SBC Communications	SBC
724 Solutions	SVNX
Sierra Wireless	SWIR
SignalSoft	SGSF
Sonera	SNRA
Sprint PCS Group	PCS
Symbol Technologies	SBL
Telecommunication Systems	TSYS
Telefónica Móviles	TEM
Telstra	TLS
Trimble	TRMB
U.S. Cellular	USM
Verisign	VRSN
Verizon Communications	VZ
Vodafone Group	VOD

Since this list is subject to change, I recommend you visit the M-Business magazine Web site, http://www.mbusinessdaily.com, for the latest list of companies in the Index and the latest stock prices. The Index serves as a useful barometer for the wireless and mobile industry and can be compared with the NYSE, Dow Jones, and NASDAQ indices.

Key Applications

Applications for M-Business within the enterprise can be broken into those that affect employees, customers, suppliers, and business partners. This short introduction to wireless applications within the enterprise will serve as a sampling of the subject matter. This matter will be treated in much greater detail throughout the rest of the book, in case studies and in the exploration of applications including business intelligence, customer relationship management, sales force automation, field force automation, and supply chain management.

Analyst surveys in the U.S. and Europe have shown that employees will benefit first from wireless enablement of the enterprise. M-Business applications will be used to increase employee productivity and will be followed by applications that are offered to customers, partners, and suppliers.

Thus, the broad categories of wireless enablement within the enterprise can be listed as follows:

>> Wireless Enablement of Employees (B2E)

>> Wireless Enablement of Customers (B2C)

>> Wireless Enablement of Partners & Suppliers (B2B)

The B2B category includes supply chain management, enterprise resource planning, and electronic marketplaces. The B2E category includes the sales force and field service workers in addition to executives, managers, and office workers.

It is important that a holistic strategy is adopted for these categories of wireless enablement. For example, employees may well need to gain access to the same applications, processes, and information as customers or partners. Additionally, as we shall discover in some of the case studies such as ADC Telecommunications, M-Business initiatives targeted for customers often turn out to be highly desirable for internal employee access as well.

All three of these categories of wireless enablement are briefly discussed in this section with a focus on some of the areas of opportunity and their benefits to the enterprise.

Wireless Enablement of Employees

Wireless enablement for employees is basically about giving employees the access to the information and transactions they need in order to perform their work-related activities. Wireless enablement can take the form of an extension of existing enterprise applications into the wireless domain. Or it can take the form of entirely new applications built from the ground up (either package or custom) specifically for use in a wireless or mobile scenario. These applications can have a profound productivity improvement for employees, the sales force, the field force, and for executives within an enterprise.

The following table (2–5) presents some of the areas of opportunity for wireless enablement of employees and the benefits that may be realized:

Table 2–5   Opportunities and Benefits for Wireless Enablement of Employees

Opportunities	Enterprise Benefits
Communications	Productivity
Basic e-mail, SMS text messaging, unified messaging, alerts and notifications	Improved employee productivity
	Improved sales force productivity
Personal Information Management (PIM)	Improved field force productivity
Calendar, contacts, tasks, to-do lists, memos	Delivery of Time-Sensitive and/or Location-Relevant Information
Intranet Access	Cost Reduction
Company directory, office locations, employee directory, hoteling/room reservations, travel arrangements, time and expenses reporting, company news	Reduced Asset Downtime
	Reduced Resource Costs (such as phone, fax, printing, mailing)
	Revenue Generation
	Increased sales

(continued)

Table 2-5 *(continued)*

Opportunities	Enterprise Benefits
Internet Access	Knowledge/Decision Making
Company Web site and applications, competitive intelligence, news	Improved Executive Reporting and Decision Making
Sales Force Applications	Delivery of Time-Sensitive and/or Location-Relevant Information
Customer/account information, product/service information, order entry and quoting, inventory management, pricing information, customer service history, lead/opportunity management, competitive information, sales reporting, training	Improved Data Capture and Accuracy
	Satisfaction
	Employee satisfaction
	Customer satisfaction
	Customer service
	Positive cultural effects
Field Force Applications	Competitive Advantage
Dispatch, project lists, service histories, inspection forms, proposals, product and part information, order processing, time and expense reporting, training	Increased competitive advantage
Enterprise Resource Planning Applications	
Executive Dashboard / Business Intelligence Applications	
Key performance indicators (KPIs), alerts and notifications, financial monitoring, operations monitoring, reporting, balanced scorecard reporting, categorized and prioritized content	

Wireless Enablement of Customers

Wireless enablement of customers can take many forms: branded cell phones or pagers to increase customer loyalty; access to hotel and airline reservations and information; telematics services for emergency location and assistance; M-Commerce transactions for wireless purchases such as stocks; wireless access to order status information; product and service information via wireless enablement of a corporate Web site; alerts and notifications on items of interest; location-based services for marketing; unified messaging for customer support; wireless gaming; informational applications such as news and weather.

The challenge on the business-to-consumer side for the enterprise is to use wireless technologies and applications in order to deepen the relationship with the customer. This needs to be done while providing applications that are easy-to-use, fulfill a need on the customer end, are actionable or informational, and support the diverse set of devices, networks, and standards in use by consumers and business customers.

Table 2–6 presents some of the areas of opportunity for wireless enablement of customers and the benefits that may be realized:

Wireless Enablement of Partners and Suppliers

Wireless enablement of partners and suppliers can take on many aspects based upon the role of the partner or supplier. Partners may be resellers, value-added resellers, distributors, wholesalers, suppliers, OEMs, industry associations, and electronic marketplace participants. A taxonomy for the various roles of partners is very much needed if one is not yet in existence. In this section, we take a look at partners in the traditional supply chain and also within electronic marketplaces.

The supply chain can benefit from wireless enablement in almost every process, including purchasing, manufacturing, distribution, and customer service and sales. Mobile technologies have long been used within the supply chain and have typically consisted of bar code scanners for improved data capture and asset management. New devices such as ruggedized handhelds from Symbol supporting both Palm OS and Windows CE operating systems, and additional means of connectivity including wireless LANs and wireless WANs mean that supply chain operations have a full range of alternatives for how, where, and when information is captured and acted upon. Information, goods, and funds flows between partners in the supply chain can now move in real-time versus near real-time or nightly batch operations.

As various industries buy and sell products and services through public or private electronic marketplaces, there is a need for continuous communications with these marketplaces in order to gather pricing information, news and events, order status, bid status, approval requests, and sales histories.

Table 2–6 Opportunities and Benefits for Wireless Enablement of Customers

Opportunities	Enterprise Benefits
Communications/Collaboration	Ability to Enter New Markets
E-Mail , SMS text messaging, unified messaging	Ability to Offer New Products and Services
Content	Increased Revenues
Advertising, loyalty/branding, travel reservations, product and service information, alerts and notifications, personal information management (PIM), location-based services, telematics, wireless gaming, remote monitoring	Reduced Costs
	Increased Customer Satisfaction
	Increased Customer Loyalty
	Competitive Advantage
Transactions/Commerce	
M-Commerce, digital wallet, preferences	
Customer Relationship Management	
Advertising, marketing campaigns, SFA functionality, order entry, order status, customer service and support, FFA functionality	

Extending these public or private electronic marketplaces with access via wireless devices can provide a solution to this need for continuous information and transactions. One of the most simple examples is that of being alerted to bidding events such as an out-bid notification during an online auction. The time criticality of the auction process means that wireless access to the marketplace is an essential tool for many participants.

The following table (2–7) presents some of the areas of opportunity for wireless enablement of partners and suppliers and the benefits that may be realized:

For more details and examples about M-Business application functionality, scenarios and benefits for the enterprise please see the M-Business Applications and Processes section within Chapter 4, Process Models and Applications for M-Business Agility. This chapter goes into detail on categories such as business intelligence, sales force automation, field force automation, customer relationship management, and supply chain management. Additionally, you can find numerous real-life case studies in Chapter 5, Industry Examples.

Table 2–7   Opportunities and Benefits for Wireless Enablement of Partners and Suppliers

Opportunities	Enterprise Benefits
Supply Chain Management	Productivity
Incident reports	Increased productivity of trading partners
Instructions and sales orders	
Just-in-time inventory management	Increased employee productivity
	Improved supplier management and productivity
Pick orders	
Delivery and receipt confirmations	Revenue Generation
	Improved partnerships
Logistics tracking	Cost Reduction
Reports and printouts	Improved quality
Quality control	Reduced inventory

(continued)

Table 2-7 *(continued)*

Opportunities	Enterprise Benefits
Vendor performance monitoring	Satisfaction
Inventory management	Higher service levels
Warehouse management	Increased customer satisfaction
Asset management	Increased customer loyalty
Mobile inventory tracking	Increased business partner
Inspections	satisfaction
Proof-of-delivery	Knowledge/Decision Making
Alerts and event notification	Improved enterprise latency
Electronic Marketplaces	Actionable information
Real-time personalized alerts based on trading events	Improved data accuracy
Order status	Increased velocity, variability and visibility within the supply chain
Collaboration with trading partners	Competitive Advantage

Business Agility Lessons

Consumer Market

>> The killer application for wireless Internet consumers varies by country, by culture, and by individual user.

>> Drivers for market adoption include industry convergence, improvements in wireless technology and standards, and cultural and regulatory forces.

>> Barriers to market adoption include business barriers related to business model and revenues, together with technology barriers related to security, product maturity, standards, usability, bandwidth, and cost.

>> Since M-Commerce is still in its infancy, it remains to be seen who the eventual winners will be. The contenders in the battle for the consumer include the wireless carriers, financial services providers, content providers, and retailers.

>> Faced with declining voice revenues, wireless carriers are looking toward value-added data services as a way to increase ARPU and to reduce churn.

Enterprise Market

>> The wireless data market has been ignited by consumers, but it is likely that the eventual winners will be enterprises that leverage the technology within their enterprise to create substantial returns on investment.

>> Carriers moving into the wireless application service provider space will need to form relationships with software companies and systems integrators and smooth internal partnerships in order to become a hit with the enterprise.

>> Business executives should ensure a strong relationship with their IT departments in order to be most successful in their M-Business initiatives.

>> The M-Business value chain is complex and involves more players than the traditional E-Business value chain.

>> M-Business can be applied within the enterprise for employees, customers, suppliers, and business partners. Employees are one of the prime targets for M-Business applications.

>> Enterprise benefits from M-Business include increased productivity, reduced cycle times, reduced costs, increased revenues, increased customer satisfaction and loyalty, competitive advantage, and improved decision-making ability.

3

Design of an M-Business

In this chapter, we look at the design of an M-Business in order to increase business agility with employees, customers, suppliers, and partners. The issues to consider are the key business needs for information and transactions with these constituencies that can be addressed by mobile business, and the current process and technology bottlenecks for delivering the required information and transactions. In addition, you should think about the timing of how these two issues, the business need for information and transactions, and the current bottlenecks, will impact the business and corporate IT. Ask yourself how your business agility is being affected by these bottlenecks today and how it may be affected in the future.

By leveraging M-Business in our interactions with our user constituencies, we increase business agility and create shareholder value in one or more of the following manners:

>> Ability to enter new markets

>> Ability to deliver new products and services

>> Increased revenues

>> Reduced costs

>> Increased productivity

>> Reduced cycle times

>> Increased customer satisfaction and loyalty

>> Increased customer acquisition and retention

>> Increased employee satisfaction

Obviously, some of these objectives are inter-connected. Increased productivity can reduce costs and reduce cycle times. The ability to deliver new products and services can enable entry into new markets, acquire new customers, increase revenues, and boost employee satisfaction. These objectives are simply listed here as high-level business strategy objectives, common among all companies, that should be taken into consideration when designing your M-Business strategy and incorporating M-Business into the fabric of your organization.

This chapter looks at the design of your M-Business strategy from a conceptual standpoint. By designing your approach with some of these concepts in mind, you will be able to engineer in the business agility that will be the enabler for your M-Business initiatives to achieve their optimal results. You can consider this chapter part of the strategic blueprint for your M-Business initiatives. In the chapters that follow, we'll look at how these concepts can be applied with a look at real applications, real processes, and real industry case studies. Additionally, we'll craft an action plan that covers M-Business strategy, architecture, and implementation.

M-Business Flexibility

The Dynamic Value Proposition

One of the first principles when designing M-Business into your business processes and applications in order to better serve business needs and remove process and technology bottlenecks, is to realize that change is inevitable. Whatever framework is developed needs to be flexible to change. The concept can be called a Dynamic Value Proposition or DVP (a term that I created in my Internet Week column when studying and commenting on the evolution of E-Business within the enterprise).

By creating a dynamic value proposition, we are meeting a business need, but also engineering in the ability for change in the information delivery, the transactional aspects of the system, and the overall business processes that are supported. As the needs of our customers, suppliers, partners, and employees change, so can our systems.

To date, business processes and their technical enablers have often had a rigid or static value proposition. A point solution is implemented and, in order to adapt to new business requirements and user needs, the entire application or process often has to be torn down and reconstructed. New requirements are collected and a new application or process is implemented. Often, by the time the new application is deployed, the business requirements have changed yet again. This cannot continue for an agile business. It creates barriers to change and is expensive in terms of resources required to deconstruct and reconstruct the value proposition being delivered. The ability to change needs to be "engineered in," so that applying change becomes less resource intensive in terms of technology, process, and human resources costs.

Much of the flexibility required in order to achieve a dynamic value proposition can be enabled with technology—especially now that we are entering the new M-Business era, where we can deliver the right value proposition to the right person at the right time and the right place.

Much of today's E-Business applications, however, need to be re-designed. They grew up during a time when time-to-market was more critical than flexibility. Large Fortune 500 enterprises scrambled to build numerous applications in order to meet business requests with little regard to the reuse of the application components or to ensure a consistent architectural framework. Even if this was desirable within the IT group, the time constraints meant there was little time to invest in creating future flexibility. We'll revisit this topic in further detail in the chapter on M-Business Architecture.

Figure 3–1 shows a matrix for measuring the dynamic value proposition of an application or initiative. In this case, it uses the variety of M-Commerce application deployment options as an example. The axes by which the dynamic value proposition is measured are the

Figure 3–1 Dynamic Value Proposition for M-Commerce Applications.

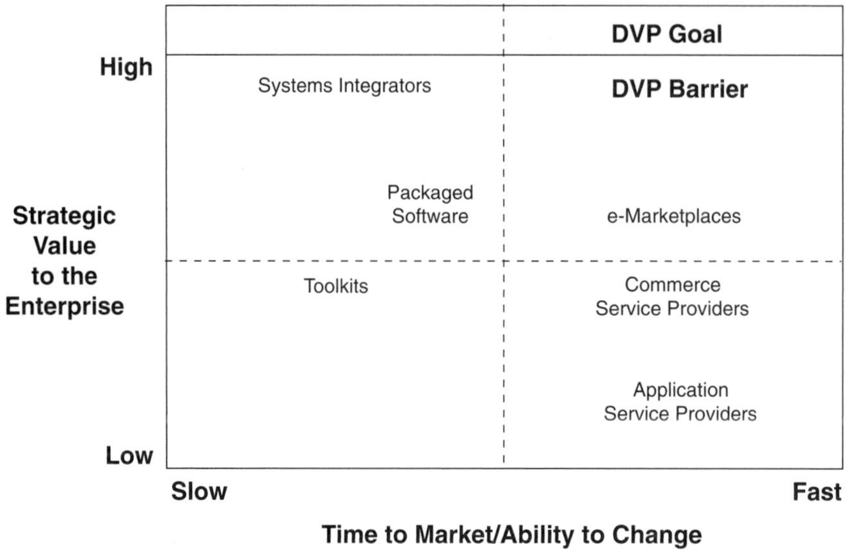

strategic value to the enterprise versus the time to market/ability to change.

The goal is to have applications and processes that have high strategic value to the enterprise, but which can also be brought to market quickly and can change quickly. The faster you can bring these strategic offerings to market, and the faster they can adapt to the changing business environment, the stronger your dynamic value proposition.

The DVP barrier in the diagram represents a transition between the static value propositions of today's applications and services and the desired goals of the dynamic value proposition applications and services of tomorrow.

If we take the variety of M-Commerce application deployment options as shown in the diagram, we can see how each offering is positioned. The M-Commerce enablers here include development environments, packaged products, and hosted services for providing M-Commerce solutions to the enterprise. We can think of some of the options as toolkits, packaged products, custom-developed applications, application service providers, commerce service providers, e-marketplaces, and so forth. All of these options are available for bringing an M-Commerce application to market for your customers and business partners.

The dynamic value proposition for M-Commerce is a system that is fast to implement (most likely outsourced), provides dynamic pricing models, dynamic commerce modes (such as auction, catalog, and reverse auction), dynamic switching between commerce modes, dynamic business rules and workflow, personalization, and support for multiple communication channels (including support for a variety of networks and device types in the wireless arena). Support for various revenue sharing arrangements and various end-user payment methods are further examples of the attributes of an M-Commerce platform that provides a dynamic value proposition.

In building these dynamic value propositions into the business, there are a many tradeoffs to be made. Outsourcing M-Business applications can create fast time to market and reduced costs, but may incur the penalty of reduced ability to change—one of the most important factors we are considering. Many outsourcers require long lead times even for simple change requests—such as a small code change—and they host generic packaged applications that vary little from one deployment to the next.

Insourcing M-Business applications may cost more and take more time to implement, but may provide an improved ability to change. The ability to change is becoming more and more possible as software components become more loosely coupled instead of hard-wired together. In this way, it may mean that the overall technology and software components used within an M-Business application become more important to the ability to change than the insourced or outsourced nature of the application.

The key takeaway is simply to think of building a dynamic value proposition into all of your M-Business initiatives. Make sure that the ability to change, in terms of both internal business rules and processes, and the ultimate value proposition that you are able to convey to your customers is available in all your enterprise systems. In this way, as your customers and your markets evolve, you can more easily adapt your value proposition to meet their needs.

The one constant in life and in business is change—ability to change, and to change rapidly, is the next competitive frontier.

In the following sections, we look at some conceptual guidelines for leveraging M-Business across your user constituencies, from employees, to customers, and to business partners.

M-Business for Employees

Focus on Activities Not Applications

Designing an M-Business for employees is often the first area targeted by the enterprise for deployment. Implementing the technology internally can yield productivity improvements, cost savings, and a relatively safe audience for pilots and experimentation. As you look at opportunities for M-Business deployment within your enterprise, it is important to understand the current processes and how wireless enablement can yield the greatest benefit. It is also important to take a user-centric and activity-centric approach versus thinking about a specific device or a specific application. As the solutions are envisioned for the user needs, so the appropriate devices, connection methods, and applications will fall out naturally.

Many of the companies discussed in the chapter on Industry Examples experienced returns on investment within 1 to 4 months by deploying wireless technologies to their workforce. However, the specifics of these applications and the connection techniques and devices utilized varied considerably from one company to the next.

In many cases, the applications arose owing to a combination of several seemingly different business drivers occurring at the same time. The Carlson Hospitality case study in our Industry Examples chapter is a case in point. The case study explores a business intelligence application for employees managing hotel properties. The initiative actually arose owing to the combination of an increase in purchase orders for PDAs together with a general request to boil down the mass of information that employees had to digest into a more usable format. The result was the Mobile Access to Carlson Hospitality or MACH-1 application that is discussed in detail later in the book.

Tradeoff Between Enterprise Standards and User Roles

Within any large enterprise there are a number of end user roles and responsibilities that are markedly different from one another. The laptop PC can often be standardized and rolled out to a large percentage of the workforce. For mobile devices, however, the standards may need to be finer-grained. For example, sales persons and executives may benefit from light-weight, sleek devices such as the Compaq iPAQ or the Palm Pilot. Field workers may benefit from sturdier, ruggedized devices from manufacturers such as Symbol. The geographic location may also change the usage behavior of these devices. Executives may tend to work in urban areas, where cellular coverage is available, whereas field workers may work in both urban and rural areas and require solutions that can function and capture data in both connected and disconnected modes.

The challenge here is to understand the tradeoff between the need to standardize on devices, operating systems, and applications and the need to give employees solutions that fit their specific work patterns and behaviors. In the end, a tradeoff will often have to be made. One device may not fit all user requirements, so the enterprise may well end up supporting both Compaq or Palm devices plus Symbol devices. However, it is often possible to standardize upon the operating system of these devices and the enterprise may be able to choose either a Microsoft Pocket PC operating environment or a Palm OS environment. Further decisions will need to be made on standards for wireless connectivity, disconnected data capture techniques, security, and so forth.

There is often no right way of doing things. Often a partially wireless solution can be just as effective or even more so than a fully wireless solution. Tradeoffs have to be made in terms of bandwidth availability and coverage while online—as in the case of a cellular call or a PDA with a wireless modem—versus the data latency issues while offline, as in the case of a PDA that is synchronized twice per day via a cradle connection to a laptop.

Focus on User Adoption and Long-Term Management

Once applications have been deployed to employees, it is important to focus on training and adoption. Some users may be using handheld computers as their first experience with computers in general. Others may be adding them to their list of devices they regularly carry around with them. To achieve the returns on investment and the productivity gains that some of the companies profiled in this book achieved, these companies were highly focused on not only getting the right solution to fit the user task, but also on getting the implementation and ongoing support processes right. They looked for vendors who not only provided robust solutions for their specific application requirements, but who also provided the tools for ongoing application management including the ability to add, change, and delete users, applications, and devices.

M-Business for Customers

When designing business agility for customers, we can leverage M-Business technologies to go a step beyond traditional personalization. We can now personalize more intelligent content for our customers—with greater time and location sensitivity built in. Along with this ability to take personalization to the next level comes a lot of challenges. We are limited by the current form factors of devices that often dictate rudimentary graphical user interfaces. We are also limited by how much customer information we can and should use.

Much has been written about customer privacy with regard to location-based services and wireless advertising. It is clear that customers should be able to dictate how and when, and even if, this information should be used. If they opt-in to receive location-based services and wireless advertising from names they know and trust, it opens the potential for totally new levels of interaction and intimacy.

A New Time-Slice for Customer Loyalty

M-Business technologies should be thought of as an additional channel for customer contact with their own set of unique characteristics and limitations. When combined with the other touch-points for communication, they provide a sum that is far greater than the individual parts. One of the main objectives with your customer initiatives should be to make it easy for the customer to do business with you—this new wireless communication channel can do just this. M-Business opens up a new time-slice in the customers' daily routine for you to continue to build and strengthen your relationship. This M-Business relationship can be applied to all the phases of traditional customer relationship management—from marketing to sales to customer service and support. You may equally well be competing for this time-slice with your competitors and new entrants to your industry wishing to acquire your customers. So there is a new touch-point and time-slice for customer loyalty, but you are competing along with everyone else for the privilege.

It is interesting to note that in addition to this new time-slice for customer loyalty, the devices the customer is using are often considered more personal than the desktop or the laptop PC. Communications and collaboration over these handheld devices such as pagers, PDAs, and WAP phones become more personal communications.

Context- and Location-Specific Loyalty

In the consumer world, this time-slice battle will most likely play itself out in a similar fashion to the wired Internet. Consumers became loyal to new Internet brands online and the top brands gained the momentum while the less effective brands often had little or no momentum in terms of revenues or site traffic. Targeting the consumer in the wireless space will be a similar battle. Only a few major brands will achieve dominance, although it will be more of a location-based dominance, or a dominance within a specific geographic area or context of activity, rather than dominance across all geographic locations and all activities. For example, the operators of wireless LANs in airports, hotels, country clubs, and train stations will have the opportunity to build the customer relationship on a location and activity-specific basis. Applications and services will be tailored to the users' environmental context and will adapt to changes in that context. These physical portals (such as airports and train stations—the real "portals" before the Internet age) will now be able to compete as virtual portals as well. They will have completely new opportunities to deliver new products and services to their customers and to interact with their customers in completely new ways.

Additionally, applications will follow the customer across his or her device types—the application or transaction can be paused on one device and resumed on another as needed. This is another feature that an enterprise should incorporate in their M-Business design—it molds or adapts the application around the customer, not vice versa.

Preference-Driven Commerce

Another concept worth consideration as you design your M-Business around your customer is the concept of preference-driven commerce— another term from my column in Internet Week. Similar to personalization, it allows you to present to the customer the relevant content and applications based upon their preferences. These preferences are pre-defined, but also configurable at run-time. Understanding the user's preferences means that you can provide the right services and information at the right time and the right location. It also has tremendous benefits for M-Business, since it allows you to transact with your customer with minimal interactions. Knowing the customer's preferences in terms of payment type, billing address, shipping address, product categories, and favorite brands can enable you to jump ahead close to the end point of the transaction. Some of the digital wallets that are being deployed today are the first signs of preference-driven commerce being applied to the M-Business arena.

If your business is consumer focused, preference-driven commerce can improve your browse-to-buy ratio. If your business is business-to-business focused, this can improve your competitive positioning by deepening customer satisfaction and making it easier for the customer to do business with you in terms of the time spent in placing orders and the time spent interacting with customer service.

The preference-driven commerce concept can also be applied to content of course. I believe it can go a lot further than the current personalized wireless portals that we see today. Some new companies are emerging (that will be covered in Chapter 9) that are taking personalization a step further toward the preference-driven content and commerce vision. There are a number of key factors that one must understand in order to move beyond basic personalization. These include the following: having a grading of interest in a category in order to enable prioritization, rather than a simple inclusive or exclusive level of interest; having the content stratified in terms of the level of detail of the information; and having this content rendered in the most effective manner based upon the characteristics of the device.

Wireless Advertising

In the business-to-consumer space, preference-driven commerce can be applied to wireless advertising. Take the case of a consumer shopping in a mall. No one wants to be bombarded by offers from every store in the mall, but if a consumer is able to specify preferences for certain product categories and certain stores and brands, the picture becomes more tangible. An offer from a preferred store with a preferred product category on sale may get the highest priority, while an offer from a preferred store with an unpreferred product category or vice versa would get a lower priority. Offers from unpreferred stores relating to unpreferred product categories would not be permitted even if they had access to the consumer location. In this hypothetical example, the only parties to have access to the consumer location would be the preferred stores, so unpreferred stores would most likely not even get a shot at the customers' business, even if they had offers of interest to the consumer. This example illustrates the importance of being on a customers' preferred list if location-based advertising becomes popular with consumers.

M-Business for Partners

When we think of the traditional value chain, we think of the relationship between the enterprise and its customers and suppliers—the flow of goods and services from their initial raw state to the final finished product or service that reaches the hands of the customer. As these linkages and relationships become more efficient and transparent via systems such as customer relationship management and supply chain management, we need to look at other enterprise relationships in order to further differentiate our business and to increase business agility.

We'll look at how M-Business can improve traditional customer relationship management and supply chain management in our next chapter, *Process Models and Applications for M-Business Agility*. In the rest of this chapter, we'll focus on a slightly different angle, that is the alliance value chain.

Figure 3–2 The Alliance Value Chain.

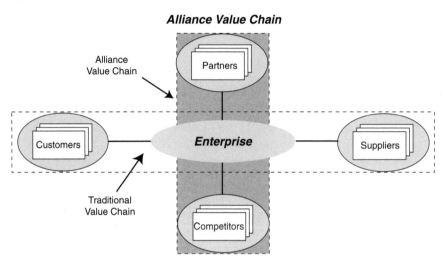

Alliance Value Chain

The Alliance Value Chain can be considered the alternate value chain from the traditional value chain or supply chain as shown in Figure 3–2 above.

The alliance value chain can be thought of as the other enterprise relationships and touchpoints that are necessary in order to help differentiate the business and increase the value proposition for customers. An example might be a software company that has partnerships with several hardware companies, networking companies, consultancies, and training companies in order to provide end customers with a turnkey solution. It is a combination of products and services from various companies that provides a better integrated solution for the customer.

The alliance value chain is most commonly related to the linkage between the enterprise and its partners. However, in some circumstances, this linkage can be extended to competitors. There are often times when competitors collaborate together on certain projects and initiatives in order to reduce costs when compared to the costs of conducting the initiative solely by themselves. A strategic alliance can also be a more favorable transaction than a merger or acquisition. It is typically far less costly, faster to pull together, and can be quickly disbanded if market conditions or the relationship dictates. In fact, Forrester Research has termed the ability to form and disband partnerships hyper-partnering—stating that

> *"To win in the accelerating digital economy, companies must form and disband partnerships faster and more often than they currently can."*

Strategic alliances are on the rise. Many of the Fortune 500 have tens, hundreds, or even a thousand-plus alliances. For example, according to a recent article in Forbes magazine, United Technologies has one hundred plus alliances, General Motors 508, Citigroup 75-plus, PBS 60-plus, Dow Chemical 130, Siebel Systems 750, Hewlett-Packard 500-plus, Cisco Systems 401, Sony 100-plus, Fidelity Investment 34, AOL Time Warner 1000-plus, and so on.

An Ecosystem of Partnerships and Alliances

BEA Systems
http://www.bea.com

In the software industry, BEA Systems stands out, along with several others including Microsoft, as a company that has built a large ecosystem of partners and has built its brand and marketshare more rapidly via this extended ecosystem. Through the BEA Star Partner Program, they have embraced a company-wide business strategy to focus on partner relationships. The program consists of an integrator track and an independent software vendor track with three levels—from one star to three star partners.

In addition, BEA has a Star Solution program that builds communities around key product, vertical, and solutions specialties with their partners. The present Star Solutions include Integration, Portal, and Wireless. The Wireless Star Solution program has nearly 25 partners who build and integrate mobile applications and services on the BEA WebLogic Server.

Overall, the Star Partner Program provides partners with relationship support, technical support, education, and technology. BEA has effectively strengthened their industry position through this ecosystem of partnerships by leveraging their alliance value chain.

Alliance Life Cycle

As partnerships are created, they typically follow a lifecycle that we shall call the Alliance Life Cycle. Alliance partners are located, due diligence is conducted on both a business and technical level, then the relationship is formally created via legal agreements and joint business plans and announced via press releases and other marketing vehicles both internally and externally. Once formed, the alliance enters the transactive stage. In this stage, joint revenues are captured and realized as customers transact with the alliance partners and consume their goods or services. Often this requires the integration of partner systems, what we may call Alliance Chain Integration.

The final stage in the lifecycle, which is an ongoing operation, is that of managing the relationship between the partners and the enterprise. This can be considered Alliance Chain Management. It encompasses all the collaboration and interaction that needs to occur both on a business level and on a technical level. For example, customer and product data may need to be exchanged, revenues tracked and shared, and physical assets and electronic assets exchanged for co-marketing and service delivery. Figure 3–3 below illustrates the alliance lifecycle and the four phases of locating, creating, transacting, and managing the relationship.

Figure 3–3   The Alliance Life Cycle.

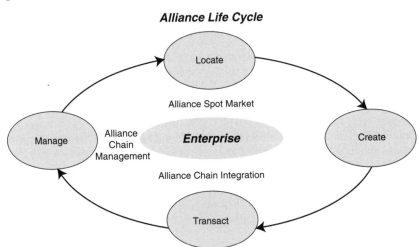

Armed with the above framework for how we might look at an alliance relationship in terms of its linkage to the enterprise and its lifecycle, we can now look at how the overall process can be improved via mobile business.

One of the areas where the concepts of the alliance value chain and the alliance life cycle are most prevalent is in industries where there is high adoption of electronic commerce. For example, electronic marketplaces can benefit from alliance relationship management solutions in order to provide value-added services to buyers and sellers within their marketplace. After a transaction has taken place, the buyer and seller may need additional secondary services such as financial services and transportation services. If a used printing press is purchased on an electronic marketplace for industrial products, the buyer and seller may need to arrange for credit check, insurance, transportation, and rigging in order to deliver the printing press from one end of the country to the other. In such cases, it may benefit the marketplace operator to form electronic relationships with these secondary, post-transaction, service providers. M-Business can play a role within these relationships by bringing information and transactions to the point of business in order to speed visibility into the alliance value chain for its various constituents.

Another example, where the alliance value chain can benefit from M-Business technologies, is the revenue-sharing that needs to occur between the wireless operators and the content and application service providers. With business models and revenue sharing still in flux, M-Business enabling these relationships can help to increase visibility into the dynamics of the M-Commerce services provided over the wireless carrier network.

Diagnosing Your Business Agility

M-Business Flexibility

>> How can M-Business enable my organization to create a more dynamic value proposition for customers, partners, and employees?

>> What attributes of my products and services need to be incorporated into this dynamic value proposition?

M-Business For Employees

>> Which employee groups stand to gain the most from the adoption of M-Business applications?

>> How can I balance the need to standardize M-Business devices, operating systems, and applications with the varied needs of my user base?

>> What usage modes make the most sense for my employees?

M-Business For Customers

>> How can I leverage the additional time-slice for customer loyalty provided via M-Business technologies?

>> How can I leverage my customers' location or context of activity in order to provide more personalized services?

>> How can I deliver enhanced customer service and support using M-Business technologies?

>> How can my customer interactions be driven via their own preferences?

>> How can I leverage M-Business technologies in order to create time- and location-sensitive sales and marketing opportunities for my customers?

>> How can I leverage M-Business technologies in order to combine the elements of online and offline customer relationship management?

M-Business for Partners

>> How many alliances does my enterprise currently have?

>> How quickly can alliances be formed and disbanded?

>> Do my current processes cover the entire alliance lifecycle?

>> How do I track and measure my alliance lifecycle?

>> How can M-Business technologies improve my alliance relationships?

4

Process Models and Applications for M-Business Agility

In this chapter, we look at the design of an M-Business from the process perspective and the realm of enterprise applications that are most applicable for process improvement. When thinking about the process perspective, one of the fundamental questions is to ask what process improvements can be made in order to improve business agility. One does this by eliminating non-value-added processes via M-Business technologies. By making these process improvements, and hence increasing business agility, we can reduce cycle times, enable real-time data and application access, enhance enterprise decision-making ability, and extend the reach of critical services to customers and employees.

In addition to improving existing processes, it is important to explore any new process scenarios made possible via M-Business. In fact, M-Business can enable us to create new processes that were previously hard to imagine. An example is the ticketless check-in and boarding process for an airplane. Using a Bluetooth-enabled device, a passenger can check-in and board a plane with no queues and with no physical tickets—the airport's Bluetooth wireless network automatically senses the passengers' presence, validates the electronic ticket, and records the boarding event at the gate. Similar scenarios are possible in many other customer-facing situations, where technology such as Bluetooth, or other wireless technologies such as wireless local area networks and wireless wide area networks, can help recognize a customer and reduce the process steps for both customers and service providers by exchanging information in both directions. The customers' personal device serves to inform the service provider or retailer of their credentials and preferences and also to ease the payment process by use of a digital wallet. The service provider can also push certain information to the customer, i.e., maps, reminders, offers and coupons, product and pricing information, and so forth.

We'll start by looking at how process can play a key role in our M-Business initiatives in terms of both opening up new process possibilities and improving existing processes. After the look at process, we will move on to explore a number of enterprise application areas for M-Business to enhance enterprise value including the following: business intelligence and executive dashboard functionality; sales force automation; field force automation; customer relationship management; and supply chain management. We'll explore each of these application areas in terms of application functionality, enterprise scenarios, and benefits.

Process Models for M-Business Agility

The Role of Process in E-Business

It is important to bear in mind the large role that process has played in the evolution and success of E-Business. One of the reasons E-Business has been so beneficial within the enterprise is due to the process improvements that were realized more than any improvements due to specific application functionality—the major gains came from automating, streamlining, and eliminating manual tasks. In many cases, E-Business applications have provided increased data accuracy through the elimination of double keying data into disparate systems or by streamlining time-intensive telephone and paper-based processes.

Enterprise data can be captured, validated upon entry, and stored in relational database management systems. Customer self-service applications often eliminate the need for employees to enter customer data completely.

A simple example is MyInsuranceInfo.com, which provides an insurance service center for end users to submit their insurance information to financial institutions requiring proof of insurance for vehicle loans. Such electronic data capture techniques mean that data is becoming more accessible, more accurate, and more useful in enterprise decision making and reporting. In summary, E-Business technologies have enabled process improvements that have led to increased productivity, reduced cycle times, and improved data quality.

M-Business Opens New Process Possibilities

There is an opportunity with M-Business to further improve enterprise processes and workflows. E-Business has re-engineered many office-based processes for manipulating information and transactions, and M-Business can be applied for the same improvements for the mobile workforce and with interactions with customers, suppliers, and business partners who are also highly mobile. Figure 4–1 shows the new communication channels and hence the new business processes and workflows that are made possible via M-Business.

Figure 4–1 New Communication Channels Enabled by M-Business.

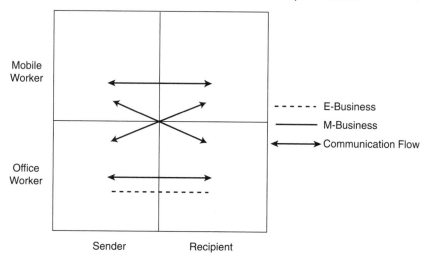

Using M-Business to Improve Existing Processes

The two most common forms of process inefficiency can be thought of as static bottlenecks and dynamic bottlenecks. If an activity is in the critical path of the overall task and involves either too much time waiting (a queue) or too much time moving around in order to accomplish a task, it can be a bottleneck for process efficiency.

Examples of static bottlenecks or queues include waiting in line at the bank, at the airport, at the rental car location, at a hotel check-in, at the amusement park, at the post office, within a store at the mall, at the car dealership, and so forth. All these examples give consumers lengthy downtime, which reduces their overall experience and takes time away from these enterprises to conduct real business with these customers. Additionally, these types of bottlenecks can also effect mobile workers, including executives and sales staff traveling on company business and experiencing the same "down time" at the airport, at the rental car location, in transit to the customer location, and at the hotel check-in.

An example of a dynamic bottleneck might be a scenario where an employee has to move from place to place in order to accomplish a task and is often forced to move away from "the point of business" in order to collect information and to enter information. Examples include field service workers in the telecom and utility industries going onto the customer premise, as well as doctors having to move around a hospital and both care for patients and frequently access patient information from back-office bound computer systems. In both of these examples, the problem is that the information systems that these workers use are not at their point of business. The field worker often leaves his or her laptop (if they have one) in their vehicle and simply takes a notepad into the customer's premises. The process is inefficient in that notes that are made on a notepad have to be keyed into the system. A similar challenge is often true for doctors. In both cases, a more efficient process is to have the information systems readily available for data capture and data reporting as close to the point of business as possible.

Applications for M-Business Agility

This chapter breaks down M-Business applications and processes into vertical applications that affect knowledge workers and executives, the sales force, field service workers, customers, and business partners in the supply chain. We also look at M-Business applications that can cross all or most of these user constituencies and are more horizontal in nature, such as business intelligence applications, productivity applications, and communication applications. Figure 4–2 presents a logical view for how we may represent the various M-Business applications within the enterprise.

Executive Dashboard and Business Intelligence

This category of M-Business application is a true horizontal. It can apply to a broad range of user constituencies across a number of different industries. The aim of an executive dashboard or business intelligence application is to deliver time-critical information to the end user related to all aspects of the business over which they have responsibility. It provides a portal-type view into the operations of the business with reports, alerts and notifications, and various forms of content. Additionally, personal information management and enterprise messaging applications may also be incorporated into the dashboard.

Figure 4–2 M-Business Applications and Processes by User Constituency.

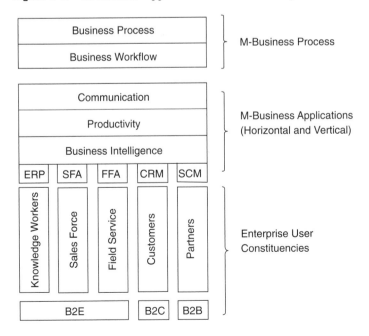

Extending the concept of the executive dashboard and business intelligence systems to the M-Business arena presents tremendous potential for exploiting the time and location sensitivity of the technology. It may often be a wearable device that executives and managers carry around with them at all times—thus enabling business-critical communication.

Functionality

Some of the functionality of the executive dashboard may include the following:

>> **Key Performance Indicators**—Monitoring and reporting on user-defined KPIs and their threshold levels.

>> **Alerts and Notifications**—Ability to send or receive alerts based upon business events. May be computer- or human-generated.

>> **Monitoring**—Similar to alerts and notifications, but usually associated with the monitoring of assets in the field such as static (meters, gauges, switches) and mobile (cars, trucks, assets) devices and equipment.

>> **Reporting**—Ability to view reports from enterprise applications.

>> **Content Services**—Feeds from various content sources such as company news, stock information, weather, travel, and so forth.

>> **Collaboration Services**—Tools for person-to-person collaboration via a variety of communications channels such as e-mail, voice, WAP push, and SMS short text-messaging.

>> **Productivity Applications**—Applications for personal information management such as calendars, address books, time and expense tracking, to-do lists, and memo pads.

A solid M-Business executive dashboard application should integrate deeply into back-end enterprise applications. As such, it will include or integrate with software components including intelligent agents, enterprise application integration technologies, enterprise portals, rules engines, workflow engines, and wireless middleware applications.

These executive dashboard applications can sit on top of wireless middleware applications. The wireless middleware technologies have made it possible for enterprise applications to be accessed via any device and any network, the M-Business executive dashboard application may now leverage that platform to build out business functionality that provides the end user with a window into key business events and metrics.

To date, many of these executive dashboard applications have been custom built. The reason for this is that their requirements can become highly specialized for the specific industry and the specific user role to which they are being applied.

For the software vendor considering building this category of product, in addition to supporting the standard laptop, PDA, pager, and WAP phone devices, these applications should also take into account in-vehicle telematics as another communication channel and potentially support a plug-and-play architecture for the wireless carriers. This way, the wireless carriers may offer the business intelligence services as part of the wireless application service provider or WASP offerings.

Scenarios

Typical applications for the executive dashboard applications within M-Business may include any of the following scenarios:

>> **Hospitality**—A hotel chain providing managers with PDAs that can monitor key business events such as changes in occupancy levels, VIP guests checking in, and key supplies running low.

>> **High Tech**—A computer manufacturer providing managers with PDAs to monitor their assigned Key Performance Indicators at all times, whether they are at their desks or are in-transit between meetings.

>> **Retail**—A retail company using PDAs to keep store managers alerted to key business events and thresholds, such as returns exceeding specified levels.

>> **Specialty Chemicals**—A sales manager using a PDA in order to monitor customer order status and receiving alerts if a major customer purchases a lower volume of chemicals for the month.

Benefits

The benefits to a mobile executive dashboard and business intelligence application are highly dependent upon the specific industry and specific user role to which they are being applied. However, general objectives and benefits that can be realized include the following:

>> **Improve Productivity**—The communications and personal information management functions within these applications allow executives and knowledge workers to remain productive even in situations where they would otherwise experience down-time, including travel to and from work (often via rail), to and from client sites, and around the corporate campus.

>> **Shorten Cycle Times**—Being alerted to key business events within the enterprise can help to shorten cycle times. The time to fix a problem or act on a key performance indicator that is out of bounds can be substantially reduced. For many heavy industries such as the energy industry and the chemicals industry, these types of notifications from equipment monitoring technologies can shorten cycle times for problem resolution and often provide substantial costs savings via reduced asset downtime.

>> **Increase Revenues**—For retailers, having the latest, up-to-date information on best-selling items from other stores and even on local weather (which may affect consumer buying patterns, particularly in the convenience store sector) can translate into increased revenues. Similarly, being alerted to a Web site inquiry from a new customer looking to purchase oil field service equipment and being able to respond immediately can translate into closing a $500K sale, whereas a similar inquiry from the same customer sits in your competitors' e-mail inbox.

>> **Reduce Costs**—Looking at business intelligence from an outbound perspective instead of an inbound perspective, one can reduce costs by automating the process of contacting customers and providing them with informational updates. Instead of having customer service representatives telephoning, faxing, and e-mailing customers, they can leverage automated enterprise messaging application platforms to perform the function for them and contact the customer via their preferred device.

>> **Improve Customer Satisfaction**—Outbound alerts and notifications can improve customer satisfaction and loyalty. They have the potential to turn a potentially negative situation such as a shipping delay or a flight cancellation into a pro-active notification which gives the customer alternate choices from which to select.

>> **Increase Competitive Advantage**—Having a business intelligence solution at hand can provide decision makers with the most immediate information on the actions of their competitors and the latest information from their own corporate intranets. Armed with this information, they can make decisions that increase their competitive advantage.

Alerts and Notifications

EnvoyWorldWide

http://www.envoyww.com

EnvoyWorldWide provides an enterprise application messaging platform for distribution of outbound messaging. The service is offered as a hosted service and allows corporations to send personalized notifications and proactive communications for time-sensitive, business-critical information to customers, employees, partners, and suppliers over a variety of communication channels including phone, fax, e-mail, SMS, WAP, PDA, and pager.

End users can specify their preferences in terms of the priority of device types upon which they would like to be notified, as well as specifying time of day and sequencing of devices. The service applies to one-to-many models, where a business event needs to be broadcast to a large user base. The business event can be either machine-generated or human-generated. Examples include notification of network outages, computer viruses, travel delays and cancellations, updated financial data, pricing corrections, shipping delays, and so on.

Customer contact information can be uploaded into EnvoyWorldWide's database, or managed programmatically through an XML-based API, prior to using the service and can also be transmitted at run-time. Administrators can then access the Web site and create notifications and track delivery via a series of Web pages. Customer notifications delivered via phone can be sent out using EnvoyWorldWide's text-to-speech technology or can be pre-recorded in the sender's voice.

Another interesting feature of the service is the ability for not only the recipient to receive communications via any device, but also for the sender to transmit communications via any device.

Process Improvement

The process improvement for the recipients of EnvoyWorldWide's service is that they are contacted immediately upon the entry of the notification into the system, either by the sender or via an automated event. Information is provided to the recipients proactively, rather than having them required to perform their own efforts in order to track the status of a particular transaction or event.

By blasting the messages to multiple devices, the end user has a good chance of being reached to receive the message immediately and in-context versus the more error-prone approach of sending a single e-mail to the user, which may not be checked until the next day, or leaving a message on the user's office phone when the user is out of the office.

The process improvement for the enterprise in using the service as the sender is that customer service representatives, and other sender roles such as field sales and service personnel, have a single mechanism for providing notifications to customers. Instead of having to try various contact methods to reach specific customers, they can rely on the automated system to cycle through the contact methods and deliver the message for them using the device preferences and priorities of the customer. This saves time for the customer service representatives and allows them to focus on other more value-added activities.

Monitoring

Notifact Corporation and US Telemetry

Notifact Corporation provides a wireless monitoring service for electromechanical equipment. One example is in the facilities management industry for the monitoring of HVAC equipment (heating, ventilation, and air conditioning). When the equipment generates an alert condition, the Notifact transmitter generates a signal via the North American Analog Mobile Phone System (AMPS) to the Notifact Web site. The transmitter is mounted on or near the equipment being monitored. The Web site is then able to send a message via a variety of communication channels (such as cell phone, pager, fax, e-mail, XML) to the appropriate recipients. The alert data can contain the error condition together with the location, make, model, and serial number of the specific piece of equipment that is at fault.

The Web site also provides a convenient location for on-going equipment monitoring for performance analysis. Customers can log-in to the site to view equipment details and heartbeat information. The heartbeat data serves to confirm that the equipment is functioning correctly.

Another company providing wireless monitoring services is US Telemetry. The company has technology that permits both static objects such as meters and gauges and mobile objects such as cars, trucks, and packages to send and receive data over their 218-222 MHz wireless network. The company provides the endpoint devices (transmitters), the cell transceiver sites, and the network operations centers for the entire solution.

Process Improvement

Note: The "monitoring" application category is often employed for field service automation, but is being included in our business intelligence category under the "horizontal" theme of alerts and notifications that can span all user constituencies.

The process improvement in the case of remote monitoring is that many steps in field service can be eliminated. For example, the schedule for field equipment inspection may be able to be lengthened from say quarterly to yearly checkups. When and if an error condition does occur, the appropriate field service technician or company is immediately notified by the equipment in question. This is far preferable than having to be discovered by the owner of the equipment and being reported into the call center. This pro-active notification can also save substantial costs by reducing equipment downtime and associated operating losses.

When communications between field service personnel and field equipment become two-way, it also opens up considerably more opportunities for process improvement and cost reduction.

Sales Force Automation

M-Business technologies can play a key role within sales force automation (SFA). Sales staff are highly mobile and spend most of their time with customers and traveling between appointments. For the salesperson in the field, there is a need for unified communications and personal information management, together with the ability to interact with corporate management and customers and to receive updates on incoming orders and order status. There is also a need for information to be provided by the salesperson to the customer in terms of product and pricing information, presentations, pre-sales support materials, scheduling of follow-up appointments and demonstrations, nondisclosure agreements, trial agreements, license agreements, contracts, quoting, order taking, and much more.

M-Business technologies can play both obvious and non-obvious roles in improving the business process for the sales force. Simple applications include the use of two-way interactive pagers for e-mail,

PDAs for personal information management (online and offline), and laptops with wireless modems for quoting and order taking. Some relatively new applications of M-Business technology include the use of cell phones as "remote control" devices in order to store and retrieve documents from the network and direct them to a fax machine.

Functionality

Some of the typical functionality found within M-Business Sales Force Automation applications is as follows:

>> **Account Information**—Information about customer accounts including company information, key contacts, purchase history, payment history, returns, service history, open customer support issues, and so forth.

>> **Product and Service Information**—Information about company products and services that can be offered to customers. This may include configuration tools when selling multipart products and services as in the high-tech industry for networking solutions.

>> **Order Entry and Quoting**—Ability to generate a quote and input an order via a wireless device.

>> **Inventory Management**—Ability to check inventory levels while at the customer site in order to minimize partial order shipments.

>> **Competitor Information**—Up-to-date information on competitors, their products, and pricing.

>> **Opportunities**—Ability to input leads, and to track opportunities through the sales cycle.

>> **Reports**—Ability to view sales reports and forecasts via a variety of filter and sort criteria.

>> **Alerts and Notifications**—Ability to send or receive alerts based upon business events such as a key customer ordering 20% less than the typical monthly order volume. Alerts and notifications may be computer-generated or human-generated.

Wireless Faxing

Thinmail
http://www.thinmail.com

Thinmail provides a network service aimed at improving the usability of e-mail between computers and wireless and thin devices. The service provides several useful features for mobile professionals such as the sales force. For example, a salesperson can use Thinmail to store product information on their network and then print that information on a fax machine by using his or her Palm VII, BlackBerry Wireless Handheld, Motorola Two-way Pager, or cell phone with e-mail.

The document is printed on a fax machine by sending an e-mail to Thinmail with instructions on the phone number to send the fax to, plus the location of the document. A typical format is as follows:

To: Fax@thinmail.com
Subject: 8005551212

- -
- John Smith, Ext 100
- Dear John, Here is my Powerpoint!
- http://www.thinmail.com/tm/v/873-f34d/killer.ppt
-

The resulting fax contains the text within the e-mail on the cover sheet, plus the full printout of the attachment listed in the e-mail.

Process Improvement

The process improvement for the sales professional is the ability to access documents on the network (in this case, the Thinmail Web-based storage) and to print them to a suitable location at or near the client site. This removes the non value-added process steps of the salesperson either calling the request in (to staff at the corporate office) or physically returning to the corporate office or other location in order to access and print out the required information. One extra process step in this new model is that the documents need to be uploaded to the Web-based storage prior to the sales visit.

>> **Collaboration Services**—Tools for person-to-person collaboration via a variety of communications channels such as e-mail, voice, WAP push, and SMS messaging.

>> **Productivity Applications**—Applications for personal information management such as calendars, address books, time and expense tracking, to-do lists, and memo pads.

Scenarios

Typical SFA scenarios where M-Business can play a key role include the following:

>> **Lead Tracking and Routing**—A sales manager at an automotive dealership can use handheld wireless devices to route customer inquiries from the Internet to the appropriate auto specialists for immediate customer follow-up. Additionally, the sales manager can evaluate the lead sources and the responsiveness of the sales teams.

>> **Customer and Competitor Intelligence**—Prior to a meeting with a key customer, a salesperson can use a laptop with a wireless modem, PDA, or WAP phone in order to access key information about this specific customer and also gain the latest intelligence on the competitors product pricing and current news releases.

>> **Collaboration**—The salesperson can also collaborate with executive management and colleagues on the sales approach and key selling messages.

>> **Document Access and Printing**—While at the customer site, the salesperson may need to access and print a document on another product for the customer to review. This may be done via a wireless printing solution.

>> **Quoting**—Additionally, the salesperson may want to provide a highly customized quote to the customer over a PDA with instant-on capability versus having to wait five to ten minutes to fire up his or her laptop and get connected to the corporate Intranet via a wireless modem or phone line—all of which are distracting to the sales process.

>> **Order Entry**—In some cases, complete order capture may be possible via the handheld device while at the customer site.

>> **Training**—Prior to a meeting with a client, the salesperson may wish to read up on the latest product information, the features and benefits of the solution, and the related services that can be provided.

Benefits

Some of the objectives and benefits of M-Business-enabled SFA applications are as follows:

>> **Improve Productivity**—With customer and product/service information readily at hand, sales staff can focus on selling rather than searching for information hidden on their laptops, or worse, back at the office. Downtime at airports and during rail travel can also be turned into greater productive time.

>> **Shorten Sales Cycles**—Sales cycles can be shortened via faster access to account information for quicker decision making and improved sales process management throughout the entire sales cycle.

>> **Increase Revenues**—With improved productivity, the sales staff can spend the time saved on closing more deals and generating more revenue.

>> **Reduce Costs**—Cost can be reduced via improved productivity of sales staff, but also via the savings on paperwork and communications that is enabled via the ability to access people and information while on the road.

>> **Improve Customer Satisfaction**—With customer and product/service information at hand, sales staff can increase customer satisfaction by knowing more about their customers and their own products and services. As such, they become more able to respond faster to answer specific questions and schedule follow-up meetings.

>> **Increase Competitive Advantage**—Sales staff equipped with real-time access to information on customers, competitors, and their own products and services are better able to take advantage of key business events that may trigger increased interest in their products and services.

Estimating Results

The return on investment for deploying M-Business extensions to sales force automation applications is highly case-sensitive—it depends upon the industry, the products or services involved, the sales process involved, the length of the sales cycle, the current amount of human and resource costs tied to closing each sale, as well as many other factors.

Productivity Improvements

If we look simply at improved productivity of the sales force when using wireless sales force automation applications, we can estimate cost savings. We can look at this from two perspectives. First, we can consider the full view of all activities where there are process improvements and generate an average productivity increase. Second, we can choose to examine activities on a case-by-case basis and calculate productivity increases for each application. The following example may help to show one of many ways to value the return on investment for wireless enablement of an individual application such as order entry and for an overall productivity increase:

Let's look at an order entry process. If a wireless application cuts down the time spent on order entry and associated paperwork from 30 minutes to 15 minutes and there are 1000 salespeople each taking 100 orders per year, we see a cost savings of $750,000 assuming an average salary of $60K per year. The formula for the cost savings for the order entry application is as follows:

Cost savings = Time savings per order (hrs) x *# orders per salesperson* x *# of salespersons* x *(avg. salary / # of hrs worked per year)*

If we take the same figures and take an overall productivity improvement of 5%, we see a cost savings of $3M. The formula for the overall productivity increase is as follows:

Cost savings = % productivity increase x *# of salespersons* x *avg. salary*

Revenue Enhancement

Of course, the main objective of deploying M-Business SFA applications to the sales force should be to maximize sales. The ROI benefits from cost savings due to increased productivity are beneficial, but the main gain is to free up more time for sales staff to do what they do best—selling. Estimating the additional volume of business that can be closed due to the increased productivity can be challenging unless we're talking about simple order taking. If we assume a productivity factor of 50% and use the data points from our productivity improvement scenario above, we find that the application has the potential to close an additional 2.5 orders per person per year or 25,000 orders in total. The formula for the overall revenue increase is as follows:

Revenue Increase = % productivity increase x *Productivity Factor* x *# of orders per salesperson* x *# of salespersons* x *avg. price per order*

Field Force Automation

The category of field force automation (FFA) is perhaps the strongest category to benefit from wireless enablement within the enterprise because the work is actually conducted in the field. In our discussion of field force automation, we will consider field workers such as field engineers, field service technicians, medical workers, government workers, and other specialized roles whose work takes place in the field. Field sales, mobile executives, and employees who have varying levels of mobility as a part of their job description are not included so our current definition for field force automation generally applies to field service workers.

Field service workers include those from many different industries. For example, we include the following industries: communications, consumer and industrial products, construction, government, health care, high tech, transportation, and utilities—to name but a few. Within the communications industry, field service technicians may service business and residential customers' equipment for voice and data services such as phone, fax, cable, and Internet. In the consumer and industrial products industry, field service technicians may service office equipment, HVAC equipment (heating, ventilation, and air conditioning equipment), oil field equipment, and manufacturing equipment. In the high tech industry, a common service is maintenance and upgrades for IT hardware.

For field workers, wireless access to information and transactions becomes a true necessity for completion of the job activities at hand. Additionally, this category is often characterized by excessive amounts of paperwork and manual processes that are ripe for process improvement. Paper-based techniques often mean that field reports and billings take days to get back to the headquarters. Both management and clients are uninformed of the work that was completed until long after the fact.

By wireless enabling the field force, management and client decisions can be made while the field force are on-site, leading to cost savings and increased billings. For example, an approval for a repair that is over the agreed upon limit can be requested and approved while the field worker is onsite, versus having to make another trip back to the site. The approval may require photographs of the work to be taken and transmitted to an off-site client or to management for approval, which explains why a simple phone call or face-to-face discussion may not prove possible. In this example, the business saves time and money by having the work completed during one visit to the site instead of a couple of visits, and the billing is increased by having the client approval to go over the agreed upon budget.

Having work history information for a specific client and a specific property or piece of equipment readily accessible at the client location may also enable the cross-selling of other maintenance activities that would otherwise have been missed. Additionally, it provides opportunities to gain increased customer satisfaction and customer loyalty. Customers often expect the same level of customer service of the field force that they expect when calling into a call center. Field workers who are knowledgeable of the customer and the service history can differentiate from the competition and provide a higher quality of service.

Functionality

Some of the typical functionality found within M-Business Field Force Automation applications is as follows:

>> **Dispatch**—Allows mobile workers to be notified of their assignments and to remain in constant communication with the corporate office.

>> **Project Lists**—Allows mobile workers to view their current assignments and those of their co-workers.

>> **Service Histories**—Access to customer and equipment service history information.

>> **Inspection Forms**—Allows mobile workers to document the results of inspections and attach digital photos where necessary. These can then be made available for clients to view on an extranet site in order to save on printing and mailing costs.

>> **Proposals**—Allows mobile workers to prepare proposals for clients to approve prior to the commencement of large work activities.

>> **Product and Part Information**—Allows mobile workers access to product and part information related to the equipment they are servicing.

>> **Order Processing**—Allows mobile workers to check part availability, to order additional parts, and to generate invoices.

>> **Time and Expense Reporting**—Allows mobile workers to complete their time and expense reports per client or per project while in the field and while the information is fresh in their minds. Everything except the physical receipts can be immediately transmitted back to the head office.

>> **Training Materials**—Access to online versions of training materials and instruction guides for equipment servicing and repair.

Scenarios

Typical FFA scenarios where M-Business can play a key role include the following:

>> **Building Repair**—Project managers within building restoration crews can use field force automation applications in order to document and bill for their work activities while onsite.

>> **IT Infrastructure**—Technicians can better service customers' IT hardware and networks via real-time access to critical service and logistics information in the field.

>> **Inspections**—Inspections of a variety of property and equipment can be conducted and documented while on-site, thus eliminating data entry requirements at the corporate office, and improving data accuracy and the speed with which decisions can be made. Applications can be as diverse as land inspections by fire departments, building and vehicle inspections by insurance adjusters, and equipment inspections by field service technicians.

>> **Law Enforcement**—Law enforcement officers can use mobile and wireless applications to provide them with up to the minute information and access to national crime databases prior to entering potentially hazardous situations.

Benefits

Some of the objectives and benefits of M-Business-enabled FFA applications are as follows:

>> **Improve Productivity**—Productivity can be improved by removing process inefficiencies in the overall administration, operations, and communications activities surrounding the field service work. This can result in improved productivity not only for the field service workers, but also for the dispatchers and administrators at the headquarters. In some cases, the human dispatch function can be completely eliminated.

>> **Shorten Service Times**—Service times can be shortened by eliminating non value-added processes, automating manual processes, and by delivering timely and accurate information to the field and improving the accuracy of data capture.

>> **Increase Revenues**—Revenues can be increased by providing field service workers with the information necessary to cross-sell other maintenance services based upon service history and also via the increase in productivity that leads to more time for billable work each day.

>> **Reduce Costs**—Costs savings can come from both the increased productivity of the field service workers and also from resource cost savings such as phone calls, faxes, printing, and mailing. Additionally, cost savings can come from reduced field worker service-related expenses due to the minimized need for return visits to fix the same problem.

>> **Improve Customer Satisfaction**—Customer satisfaction can be improved by the more professional image conveyed by workers that know more about the customer and the equipment service history and are able to resolve more questions and issues that occur in the field.

>> **Increase Competitive Advantage**—In an industry role that has long been seen as low-tech and as a service nightmare for customers (workers showing up without the required parts to fix the problem, not knowing any of the prior service history, or not even understanding the task required by the customer), the wireless enablement

of field service applications can create an immediate differentiator for those companies that make it a part of their strategy.

Estimating Results

As in the sales force automation section, the return on investment for deploying M-Business extensions to field force automation applications is again highly case-sensitive. However, the following example may help to show one of many ways to value such initiatives.

If we look simply at improved productivity of the field force when using wireless field force automation applications, we can estimate cost savings. If the application saves just two hours per week in productivity for a field worker and there are 100 workers each getting paid $50 per hour, we see a cost savings of $520,000. The formula for this productivity increase is simply as follows:

Cost Savings = Time Savings (hrs/wk) x *52* x *Avg. Salary* x *Num. of Field Workers*

With the current inefficiencies in the field service process, there is tremendous potential to increase productivity. Typical productivity increases will be between 5% and 15%, but if processes are currently highly manual the increase can go as high as 30% or more.

When factoring in the improved productivity of both field service workers, dispatchers and administrative staff, the new amount of time opened up for additional billing, and the resource cost savings, the return on investment for an M-Business FFA implementation can be achieved within a matter of a few months.

Customer Relationship Management

Customer relationship management (CRM) is a large category that actually encompasses sales force automation that we covered earlier in this chapter. In addition to sales force automation, CRM includes marketing automation, customer service and support, and any other applications or processes that have touch-points with the end customer. The goal of CRM is to increase shareholder value through better understanding the needs of customers and adapting to those needs through new and improved customer relationships.

M-Business CRM for Customers

Just as consumers and business customers started to adopt the Internet as a way to obtain product and pricing information, order status information, and customer service, they will also adopt and expect to be able to use the M-Business channel for the same kinds of information and interaction. They may also expect this channel to provide them with a faster response and with greater relevancy to their current location or situation.

A lot of customer relationship management is simply about giving customers the information and the applications they need in order to help themselves. They may require access to the same, or at least similar, kinds of product and pricing information that you provide to your sales force or to similar kinds of product documentation and service information that you provide to your field force. Equipping the customer with the tools to help themselves can help the enterprise reduce costs and can also result in greater customer satisfaction. Done incorrectly, however, it has the potential to create the opposite effect.

Increasing Complexity of Customer Interactions

An additional channel for customer communications such as the M-Business channel (typically wireless access via cell phone, PDA, or pager) actually means more complexity for your enterprise. Customers will soon come to expect the multi-channel support enabled via M-Business; the challenge for the enterprise will be to manage this complexity and ensure that all touch-points work successfully and correctly for customer transactions.

Ideally, all communication touch-points feed into a central customer-centric view of the world, which your customer service staff have access to. Changing the "rules" between communications channels can create problems both for customers and customer service staff alike. It can have the effect of departmentalizing the various communications touch-points so that customers are routed to different call centers based upon their incoming communications channel. You may have experienced this yourself when perhaps purchasing tickets for air travel and having different numbers to call for telephone booking versus "Web" booking when the Web site is not functioning properly. The situation is often aggravated by different ticket pricing over the phone versus over the Web, with telephone call center agents unable to provide the $100 cheaper tickets that are quoted on a Web site that will not complete the final booking.

Simply stated, M-Business will challenge customer relationship management systems. They will have to not only adapt the presentation layer to present content in the appropriate manner for a variety of devices, but these systems will also have to become more intelligent. They will have to know exactly what information to collect from the customer and understand all the permutations about the information and transactional requests being conducted. There is still a tremendous way to go here. Even Web applications have a long way to go. To give an example, how many airline Web applications can allow you to book travel that is made up of part frequent flyer mileage and part paid travel? These simple scenarios often mean the customer is forced to resort to regular telephone calls with an agent or agents.

Providing Intelligent Customer Service

Despite the challenges of adding further intelligence into customer relationship management applications, especially those enabled via M-Business, the opportunities for increased customer satisfaction, increased revenues, and reduced costs are tremendous.

M-Business technologies provide a way for the enterprise to take a more proactive stance regarding the customer relationships. Instead of waiting for a call to customer sales or service and telling the customer what they cannot do, the enterprise can now contact the customer and tell them what they can do. M-Business-enabled CRM is all about becoming more pro-active in your communications with your customers and often turning customer service and support interactions into new ways to deepen the customer relationship. Moreover, it allows one to find new sales opportunities via personalized, time, and location-aware interactions.

Intelligent M-Business devices may well be able to communicate not only with the customer, but with his or her surroundings and his or her surrounding products. Thus, an intelligent device with a software agent running on behalf of a manufacturing company could detect the devices within the personal area network of the customer, perform a health-check on the devices, and contact customer support if anything is detected to be out of the specified tolerance limits. This scenario takes remote monitoring a step further by immediately bringing customer support online via the customers' PDA; for example, as soon as there is an anomaly with the equipment. This hypothetical example may one day be an example of world-class customer service—leveraging M-Business technology in order to detect a problem before the customer is aware of the problem and immediately launching a rich media communication channel such as a live video link in order to solve the problem.

M-Business CRM for Employees

The discussion so far has focused largely on customer self-service, but M-Business-enabled CRM applications can also be leveraged by employees engaged in marketing, sales, and service activities. The technology can provide up-to-the-minute information on marketing campaigns, sales activities, and service levels for executives, managers, and marketers who are frequently away from the office. The applications can provide both analytical and operational CRM data and allow employees to access and react to customer-centric information no matter where their location or the location of their customer happens to be. For example, a marketer may be able to access real-time information from a marketing campaign and be able to react accordingly. A customer service representative may be able to access real-time customer service level information and be able to react to problems before they start to escalate.

Functionality

Some of the typical functionality found within M-Business Customer Relationship Management applications is as follows:

Marketing

>> **Advertising**—Ability to send out personalized advertisements to customers based on their location. Ability for customers to receive and act on these ads while mobile.

>> **Marketing Campaigns**—Ability to conduct and analyze marketing campaigns while out of the office. Ability for customers to receive and respond to marketing campaigns such as surveys or offers while mobile.

Sales

>> **SFA Functionality**—Including account management, sales management, lead management, time management, and contact management.

>> **Order Entry**—Ability to enter customer orders while at the customer site. Ability for customers to make purchases while mobile.

>> **Order Status**—Ability to check on order status while at the customer site. Ability for customers to check order status while mobile.

Service

>> **Customer Service and Support**—Ability to deliver customer service and support via wireless devices. Including incident assignment, escalation, tracking, and reporting.

>> **FFA Functionality**—Including call scheduling, problem resolution and knowledge bases.

Scenarios

Typical CRM scenarios where M-Business can play a key role include the following:

>> **Marketing**—Providing customers with targeted advertising and offers based upon their location or their personal profile. Use of over-the-air provisioning in order to immediately provision content and applications based upon the users' response. For example, providing an offer to a wireless subscriber to evaluate a time and expense reporting application for small business.

>> **Alerts and Notifications**—Providing customers with alerts and notifications to pagers, PDAs, and cell phones that enable them to be informed of key events and to take suitable action. For example, providing flight cancellation information and allowing customers to take suitable action such as booking on the next available flight.

>> **Contacting Customer Support**—Use of two-way interactive pagers to contact and communicate with customer support.

>> **Linking Remote Monitoring With Immediate Customer Support**—
Use of a PDA to communicate and diagnose equipment within the
customers' personal area network. Launching a video session with
a live customer support representative if any equipment anomalies
are detected. Of course, this is a future scenario, but one that is certainly possible.

>> **Accessing Customer Support Materials**—Providing customers
with the ability to access product information and documentation
to help them troubleshoot various problems in the field. For
example, providing an oil and gas operator with information
required in order to fix a problem with a pump at the well-head.

Benefits

Some of the objectives and benefits of M-Business-enabled CRM
applications are as follows:

>> **Increased Customer Loyalty**—Customers who can reach your
organization wherever and whenever they need to are likely to
become more loyal than to those organizations who do not provide this service. M-Business-enabled CRM will soon become a
matter of convenience. The benefit for organizations who provide
this extra level of service will be increased customer satisfaction
and loyalty leading to increased revenues and reduced customer
churn.

>> **Improved Sales and Marketing**—M-Business-enabled CRM pro-
vides a way for your organization to tie together both online and
offline sales and marketing approaches. Offers delivered via wireless devices based upon customer location can immediately be
redeemed within physical stores. Additionally, non-location
dependent offers and shopping lists can be stored on handheld,
wireless devices for redemption within physical stores such as grocery stores.

>> **Improved Customer Service and Support**—Customers can be pro-
vided with improved service and support not only on a 7 x 24
basis, as was enabled by the Web-based services, but also on a
location-dependent basis via wireless devices.

>> **Improved Productivity and Cost Reduction**—With access to customer data and reports available to executives, managers, and marketers via wireless devices, organizations can experience improved productivity and cost reduction by eliminating non value-added steps in locating and accessing customer reports and data.

Supply Chain Management

The traditional supply chain consists of a linear flow of goods from suppliers to manufacturers to wholesale distributors to retailers and finally to customers. The E-Business supply chain has increased efficiencies and visibility throughout the supply chain, but has often increased the number of participants and interactions occurring within the supply chain. For example, in addition to the traditional participants, there are now private and public electronic marketplaces providing informational and transactional services. These services include the primary transactions of buying and selling the goods themselves, plus the secondary transactions of buying and selling additional services such as financial services and logistics services. Additionally, as companies have focused on their core competencies and have outsourced other non-strategic processes, other service providers have been brought into the equation.

Unless managed properly, the resulting increase in information and transaction flows and the increased number of participants can actually hurt supply chain efficiencies. M-Business-enabled supply chains provide a way to improve the business agility within the system. Speed and flexibility can be improved in terms of the speed at which information flows through the supply chain across partners and suppliers. M-Business-enabled supply chains also improve the flexibility with which these partners and suppliers can react and can even be interchanged based upon performance.

There are applications for M-Business enablement of the supply chain in almost all process areas including purchasing, manufacturing, distribution, and customer service. It can speed information flows, product flows, and funds flows across all participants. Perhaps the most important factor related to M-Business enabled supply chains is that they can increase data accuracy and speed decision making.

"Wireless" Versus "Mobile" Applications in the Supply Chain

Many of the current solutions implemented within the enterprise today are mobile solutions that provide the required business benefits without the full implementation of truly constant wireless connectivity. They are often designed for a combination of online and offline usage and have varying degrees of wireless access.

For example, many enterprise operations and supply chain scenarios employ devices such as the Symbol SPT 1500 handheld, based on the Palm OS, for asset tracking, inventory management, and physical data collection. The device is basically a ruggedized Palm III with bar code scanning technology incorporated. These devices store the data locally on the handheld device and then allow data to be uploaded to the back office systems via the Palm HotSync technology. If the data is synchronized via a cradle, the argument can be made that this is not a true "wireless" solution. The answer is that while some enterprise applications are mobile solutions to the business problem at hand, and while others are wireless LAN or wireless WAN solutions, they all achieve the desired effect of increasing business value through M-Business. Some solutions achieve their desired results via mobility and wired connections, whereas others achieve their results via mobility and a combination of full-time or part-time wireless access.

The decision as to which mobile technologies to use should be made based upon the business issue at hand, the required or acceptable data latency in both upload and download directions, and the location of the enterprise operations and workforce. If nightly data transfers to the corporate office are acceptable and the information residing on the device such as inspection forms is fairly static, then a cradle solution may be appropriate. If real-time access to information is required, then an available wireless connection over a wireless LAN or a wireless WAN is required. To conclude the Symbol example above, the company makes additional devices to support these scenarios via their SPT 1700 with wireless LAN support and their SPT 1733 with wireless WAN support. These are just a couple of examples of the types of devices on the market for supply chain operations that require rugged handheld computers. Additionally, Symbol makes devices based on the Windows CE platform and a variety of products including key/pen/touch computers, handheld and wearable scanning devices, stationary and vehicle mount terminals, and wireless LAN products.

Functionality

Some of the typical functionality found within M-Business Supply Chain Management applications is as follows:

>> **Incident Reports**—Ability for employees to document incident reports such as supplier product quality issues.

>> **Instructions and Sales Orders**—Employees can access instructions on where to store goods, how to handle them, sales orders for outgoing transports, stock transfer orders, and so forth.

>> **Inventory**—Ability to employees to check warehouse inventory.

>> **Just-In-Time Inventory Management**—Ability to enter maintenance data on equipment in order to improve maintenance planning.

>> **Pick Orders**—Pick orders can be transmitted to employees working in the warehouse on outgoing shipments.

>> **Delivery and Receipt Confirmations**—Employees can use handheld scanners to scan barcodes on incoming and outgoing pallets.

>> **Logistics Tracking**—Containers can transmit location information for shipments in transit.

>> **Reports and Printouts**—Employees can get reports and printouts on supply chain data. Customers can gain access to inventory lists and reports on goods movement.

Scenarios

Typical SCM scenarios where M-Business can play a key role include the following:

>> **Quality Control**—A retailer using a handheld solution in order to improve the inspection process for merchandise received from vendors prior to placement on the shop floor. By recording defect information at the point of delivery, the company can shorten the time delay between problem detection and vendor resolution and also reassign back-office data entry functions to more productive work.

>> **Vendor Performance Monitoring**—A postal company using wireless devices to document incident reports related to mishandled or damaged mail receptacles as a result of airline transportation.

Issues are more accurately and more rapidly documented versus paper-based techniques and possible airline changes or fines can be made due to the improved vendor monitoring.

>> **Warehouse Management**—Mobile data entry allowing paperless picking for a beverage distributor. Wireless terminals can transmit information captured during the beverage picking or storage process into an ERP system. This ERP system can then communicate directly with customer systems via EDI.

>> **Inventory Management**—Mobile data entry allowing a postal company to improve reporting methods in order to analyze automatic letter sorting machine failures more accurately. Increased information about failures improves maintenance planning and enables reduced downtime and lower inventory costs.

>> **Asset Management and Mobile Inventory Tracking**—Ability for technicians in the aviation industry to scan equipment barcodes with handheld devices to retrieve and update inventory information for equipment that is frequently relocated and reassigned.

>> **Inspections**—Ability for inspectors working for an automotive manufacturer to inspect incoming vehicle shipments using handheld devices as they arrive at various ports. Handheld data capture with nightly synchronization to a central location provides faster access to inspection data and allows managers to update inspections forms based upon particular makes or models of vehicle.

Benefits

Some of the objectives and benefits of M-Business-enabled SCM applications are as follows:

>> **Improved Customer Service**—Integration of the warehouse with customer systems can improve customer service levels.

>> **Productivity Enhancements and Cost Reduction**—Many supply chain metrics can be optimized including delivery performance, inventory reduction, fulfillment cycle time, forecast accuracy, overall productivity, lower supply chain costs, fill rates, reduced equipment downtime, and improved capacity realization.

>> **Improved Supply Chain Visibility**—Greater visibility into supply chain information flows, product flows, and funds flows between companies.

>> **Improved Supplier Management and Product Quality**—Improved ability to inspect and monitor the quality of supplier products and services as they are delivered.

>> **Improved Data Accuracy**—Data collected via wireless devices via up-to-the-minute collection forms and validation rules loaded on the devices can be more effectively collected and validated.

>> **Reduced Cycle Times**—Reduced cycle times for information processing due to automated data capture and validation compared to manual techniques.

>> **Faster Decision Making**—With data captured at the point of activity and transmitted wirelessly to enterprise systems and customers, decisions can be made on the spot versus waiting days or weeks for paperwork to be completed and submitted.

Business Agility Lessons

M-Business Process

>> M-Business presents an opportunity to further improve enterprise processes and workflows.

>> M-Business technologies present us with the opportunity to completely re-engineer old processes and create new processes that were previously hard to imagine.

>> One of the fundamental questions is to ask what process improvements can be made in order to improve business agility by eliminating non value-added processes via M-Business technologies.

>> The two most common forms of process inefficiency can be thought of as static bottlenecks (queues) and dynamic bottlenecks (information away from the point of activity).

Diagnosing Your Business Agility

Executive Dashboard/Business Intelligence

>> Does my business have key performance indicators that need to be monitored in real-time by employees?

>> What areas of the business such as finance and operations can benefit most from real-time monitoring and alerts based upon these key performance indicators?

>> How can these key performance indicators be applied to align employees day to day activities with the business strategy?

Sales Force Automation

>> How can my sales force beat the competition via real-time competitive intelligence?

>> What customer and account information will help my sales force become more productive?

>> What product and service information will help my sales force in cross-selling and up-selling?

>> What lead management and sales reports would benefit my sales force while on the road?

Field Force Automation

>> How can field force productivity be improved via M-Business applications?

>> What information does my field force require in order to be more effective?

>> How can my field force generate additional revenues via access to customer preferences and service history?

Customer Relationship Management

>> How can customer loyalty be increased via M-Business?

>> How can customer service and support be provided by M-Business applications?

>> How can sales and marketing benefit from M-Business applications?

Supply Chain Management

>> Which areas of my supply chain can benefit from data capture at the point of activity?

>> Which connection techniques such as cradle, wireless LAN, and wireless WAN are most appropriate for these areas?

>> How can M-Business applications improve customer satisfaction and vendor management with real-time visibility into the supply chain?

5

Industry Examples

In this chapter, we examine several examples where early adopters have implemented M-Business applications within their enterprise and have achieved significant and demonstrable results. The chapter is grouped by industry for easy reference. I have divided the industry categories into the following broad classifications:

>> Communications and Content

>> High Tech

>> Consumer and Industrial Products

>> Public Services

>> Health Care

>> Financial Services

Within these core six verticals there are important sub-classifications that most often define entire industries. For example, financial services can be considered as comprising banking, insurance, and real estate and hospitality. Communications and content may include the communications companies such as the wired telecommunications companies, wireless carriers, and next generation communications service providers, as well as content companies such as media and entertainment companies. In the consumer and industrial products vertical, one can include the retail, industrial products, automotive, and travel and transportation industries. The industrial products sector may also include the energy and mining industry and the specialty chemicals industry to name just a few. Within public services, we can include a number of government institutions and universities. There are many ways to classify these industries and this is a high-level classification that aims simply to place the case studies in appropriate sections.

The following examples aim to demonstrate how true business value has been realized from M-Business technology across this entire spectrum of industries. When pulling together these case studies and success stories, my focus was to look for interesting established and upcoming software companies and technology "firsts." More importantly, these examples serve to showcase large enterprises that had implemented M-Business and had seen measurable results that made a difference to their business and to their customers.

Communications and Content

Within the communications and content industry, there are two main categories of company. The first category are those companies that provide communications services such as local exchange carriers, wireless carriers, long distance carriers, Internet service providers, application service providers, and next generation communications services providers. The second category are those companies that provide content, such as media companies and entertainment companies.

Since both of these categories are a core part of the wireless Internet value chain, they are able to exploit M-Business technologies not only internally, in order to benefit the bottom line through increased productivity, but also in the form of new products and services that can directly impact top-line revenues.

For media and entertainment companies, M-Business presents an additional channel over which they can deliver their content. It presents an opportunity to connect with customers in new ways on their own terms—providing the right content at the right time and the right place. Over-the-air provisioning, where the consumer can gain immediate access to new applications and services for a fee, means that these companies can use the wireless channel not only as an additional touch-point for building the customer relationship, but also as a new revenue channel.

We touched upon the opportunities for the telecommunications companies earlier in the book. Communications companies can literally transform their business models by moving into value-added data services in order to increase average revenue per user (ARPU) and to reduce churn. Companies such as NTT DoCoMo have become models for the wireless industry having successfully deployed their data services to their subscriber base with high penetration rates and high revenues.

Improving Employee Productivity and Customer Service With Wireless Order Status

ADC Telecommunications
http://www.adc.com

Executive Summary

ADC Telecommunications is a network equipment manufacturer based in Minnetonka, Minnesota. The company employs over 19,000 people worldwide and has annual sales of over $3.3 billion. They deployed a wireless order status application for customers and employees and achieved an ROI in less than 1 month. They saved 450,000 man-hours per year for 2,100 employees. On average, their employees saved 2 to 5 hours per week by using the application.

Company

ADC Telecommunications

Name: ADC Telecommunications

Web Site: http://www.adc.com

Symbol: ADCT (NASDAQ)

Business: Network Equipment

HQ: Minnetonka, MN

Employees: 19,000 worldwide

Revenues: Annual sales of more than $3.3 billion

Solution

Category: CRM and FFA

Application: Wireless access to order status information

>> Order Status
>> Shipping Dates
>> Tracking Numbers

Technology: Air2Web hosted service

Target Audience: Customers and Employees

>> 1,100 installers
>> 130 call center agents
>> 900 account managers

Devices:

>> 2000 Palm Pilots for Employees
>> WAP, SMS, and Palm VII for Customers

Challenge and Business Drivers

Business Drivers

>> Cost Savings

>> Competitive Advantage

Former Process

>> Installers used notebook computers and phone jacks to look up order status information

Benefits

Time Savings

>> Installers—5 hours per week

>> Call Center Agents—5 hours per week
>> Account Managers—2 hours per week
>> Total of 450,000 man-hours per year for 2,100 employees

ROI

>> Less than 1 month

Challenge

ADC Telecommunications wanted to leverage wireless Internet technologies in order to generate cost savings and also to maintain competitive advantage. According to Theresa Enebo, Manager of Client Systems and Services, ADC started the project as a way to get their feet wet with wireless technologies and to gain some early experience and competitive advantage.

They focused on a single application for their initial wireless deployment: order status for customers and employees. Although the initial focus was on the customers, the application became popular internally and ended up being both employee and customer focused. The current process for employees to access order status information was time consuming with installers using notebook computers and having to search for phone jacks while on the customer premise.

Solution

The solution ADC chose was the Air2Web hosted service. The implementation took approximately four months, which included requirements analysis, coding, and testing. The order status application was rolled out to over 2,100 employees, including installers, call center agents, and account managers. The initial audience for the rollout was a couple of dozen employees assigned to their two biggest customer accounts.

Palm Pilots were purchased for the employees, along with leverage of existing employee devices such as Palms and SMS-capable cell phones. The solution was also designed in order to support customer access via SMS, WAP, and the Palm VII. ADC found that two-thirds of their employees used their existing SMS-capable phones to access the application and the remaining one-third used Palm VII devices with access via Palm.Net.

SMS was a key access method because it was a lowest common denominator with many employees. Moreover customers already possessed the devices. Using their SMS phones, users were able to obtain order status information either on-demand or via notification. The on-demand mechanism works as follows: The user dials a pre-determined phone number, the system recognizes them via caller id, and after they hang up, it sends the information to them via SMS. For users who wanted to enter a custom order number, an IVR system provided similar functionality but allowed them to key in the order numbers.

Information accessible via the application included order status information, shipping dates, and tracking numbers. In the future, ADC plans to add further functionality such as product pricing information, product availability, and ADC contact information such as sales contacts and office locations. An additional feature is that ADC plans to link in the UPS wireless tracking functionality so that customers can obtain full order status information and tracking in a single session versus having to visit two separate sites over their wireless devices. UPS is also a customer of Air2Web.

For training purposes, ADC used Webex in order to demonstrate the application. While they implemented support processes for customers and employees using the application, they found that they had very limited call volume from these end users, which indicated that the application was a success from a usability standpoint. Interestingly, one of the first challenges they faced when planning the solution was to convince themselves that the application would indeed be usable given the subset of order status information they were delivering when compared to the complex order status application they currently supported on their Web site. By remaining open-minded about the wireless version of the application and not focusing too heavily on the technical differences between Web and wireless implementations, they were able to deliver a wireless solution rapidly and experience very positive feedback from end users.

Benefits

The benefits of the deployment of the wireless order status application were considerable. The return on investment was generated in less than 1 month. 450,000 man-hours per year were projected to be saved when looking across the 2,100 person user-base. ADC estimated that between 2 and 5 hours per week were saved by using the application when compared to the prior order status process. These numbers were determined by interviewing end users and finding out how much time they were spending on order status determination. This involved both phone calls to customers and phone calls to each other—the calls were typically placed between the account manager and the customer service representatives.

The 450,000 man-hours per year figure was determined by taking the number of employees using the application and multiplying by the average figure for time savings based upon interviews.

By taking a simple application and transforming how it was delivered to both customers and employees, ADC was able to experience a substantial return on investment and increased customer satisfaction.

High Tech

In the high tech industry, companies can leverage M-Business to improve their supply chain efficiencies, operate more efficient electronic marketplaces, and to increase customer satisfaction and loyalty. Companies within this industry either offer technology products and services or use technology as a pure-play business model, as in the case of the business-to-business electronic marketplaces. In many cases, these companies are a critical part of the wireless Internet value chain, providing the hardware, software, and networking components we need to make M-Business a reality. In addition to being on the supply side of the equation, these companies can also leverage

M-Business as an enterprise themselves in all the areas we have discussed including customer relationship management, sales force automation, field force automation, supply chain management, and enterprise resource planning.

We'll now take a look at eBay as an example of a high tech company operating an electronic marketplace and enhancing its global reach and customer satisfaction and loyalty via M-Business technologies. Bear in mind that this example can also be extended to the business-to-business electronic marketplaces that facilitate the sale and purchase of high tech equipment such as electronics, computers, and networking equipment. Wireless middleware companies such as Brience have done just this by helping companies such as Ingram Micro to wirelessly-enable their electronic procurement hubs.

Enhancing Global Reach and Customer Satisfaction and Loyalty

eBay
http://www.ebay.com

Executive Summary

eBay wanted to increase their global reach with customers and to provide them with anywhere, anytime access to their online auction service. They implemented a wireless solution using technology from 2Roam. The wireless functionality includes the ability to register with eBay, to access personalized content on the My eBay site, to search for items of interest, to view featured auctions, to browse by category, to access the latest bidding information, and to place new bids—essentially everything that an eBay buyer or seller would wish to perform on the site. The wireless application supports a variety of device types and standards including WAP, CE, Palm, RIM, and SMS. The benefits for eBay and their customers have been greater global reach for the marketplace, more efficient marketplace functionality, and increased customer satisfaction and loyalty.

Company

eBay, Inc.

Name: eBay Inc.

Web Site: http://www.ebay.com

Symbol: EBAY (NASDAQ)

Business: Online Auction Service

HQ: San Jose, CA, U.S.A.

Employees: 1,927 (2000)

Revenues: $431 million (2000)

Solution

Category: CRM

Application: eBay Anywhere

>> Registration
>> My eBay personalized content
>> Search
>> Bidding

Technology: 2Roam

>> Delivered eBay's wireless site to the specification of Sprint PCS and AT&T Wireless portals

Target Audience: eBay Customers

>> eBay country-specific sites in the U.S., U.K., Canada, France, Germany, Japan, Australia, Italy, and Korea with users representing 150 different countries

Devices:

>> Range of consumer devices including WAP, CE, Palm, RIM, and SMS

Challenge and Business Drivers

Business Drivers

>> Increased global reach
>> Increased customer satisfaction and loyalty

>> Increased customer audience

>> Anywhere, anytime service

>> More efficient marketplace functionality

Former Process

>> Web site presence with limited wireless access for customers

Benefits

Increased Customer Satisfaction and Loyalty

>> Anytime, anywhere access to eBay from any wireless device

>> Increased global reach

>> More efficient marketplace functionality

Challenge

eBay is the world's online marketplace with country-specific sites including Australia, Canada, France, Germany, Italy, Japan, Korea, United Kingdom, United States, and users representing 150 different countries. eBay identified wireless users as their next growth area and sought to increase customer loyalty and reach by offering enhanced convenience through a wireless solution. Since in many international markets wireless penetration exceeds PC penetration, and many people access the Internet for the first time over wireless devices, this new channel was an important way to extend eBay's global reach. They also wanted to make their marketplace more efficient by allowing anywhere, anytime access from a variety of wireless devices.

Solution

eBay was a very early adopter of wireless technologies and made a strategic partnership with 2Roam back in May, 2000. The 2Roam partnership enabled eBay to wirelessly enable their site and to support the different design requirements of both the Sprint PCS and AT&T Wireless portals that feature the eBay site. 2Roam is a software company based in Redwood City, California that provides a wireless middleware platform as either a fully hosted ASP solution or as a licensed in-house server solution. The eBay solution runs on 2Roam's hosted ASP solution.

The wireless functionality includes the following: the ability to register with eBay; to access personalized content on the My eBay site, so that users can view the progress of their current auction bids and view their feedback; to search for items of interest by item number or keyword; to view featured auctions;to browse by item category and sub-category; to access the latest bidding information; and to place new bids. One can do essentially everything that an eBay buyer or seller would wish to perform on the site other than the posting of items for sale. Posting an item for sale is something that is best performed over a wired connection owing to the amount of text that needs to be entered relating to the item.

According to Todd Madeiros, Director of International Marketing and eBay anywhere, "eBay has been able to control its brand identity and optimize look and feel across many countries, technologies, and devices." The eBay wireless application supports a variety of device types and standards including WAP, CE, Palm, RIM, and SMS.

Benefits

The benefits for eBay and their customers have been greater global reach for the marketplace, more efficient marketplace functionality, and increased customer satisfaction and loyalty. With the time-criticality of online auctions, the eBay anywhere service has enabled participants to monitor their bidding status and place new bids while mobile. This effectively empowers those with wireless devices to have a greater level of access to the marketplace and control over their online auctions.

Consumer and Industrial Products

The consumer and industrial products industry covers a large number of industry verticals including retail, industrial products, automotive, and travel and transportation industries.

The examples that I have chosen to include as case studies include Aviall, Office Depot, and Rental Service Corporation. The Aviall case study shows how wireless technologies can be used to create a supply chain management solution that improves customer service and reduces costs. The Office Depot case study shows how wireless technologies can be used to implement a delivery management system that benefits both employees and customers by providing real-time access and visibility into the delivery process and reducing costs. The Rental Service Corporation case study shows how wireless technologies can be used to implement a sales force automation solution that empowers the sales force not only to be more productive in terms of account management, but also to generate more revenues via improved territory and account management.

Improving Customer Service Via a Wireless Aviation Parts Supply Chain Solution

Aviall

http://www.aviall.com

Executive Summary

Aviall, Inc. is the world's largest independent distributor of new aviation parts and a leading provider of inventory information services. The company represents some 180 quality manufacturers and distributes and markets parts to more than 13,000 general aviation operators and 300 airlines worldwide.

For key customers, Aviall provided a FreeStock program where they performed monthly onsite inspections of part quantities at customer warehouses and replenished their inventory where necessary. This process was time-consuming and manual. Aviall wanted a way to increase their level of service for these customers, improve data accuracy while performing inventories, and to shorten the cycle times in order to deliver the service to additional customers.

They implemented a wireless supply chain management solution. They automated the process using software from CellExchange together with Symbol SPT 1700 handheld devices in order to better serve their customers.

They estimate a cost savings of over $1 million owing to reduced time required to perform these parts inventories at customer warehouses. Additionally, the solution has enabled them to increase customer satisfaction, improve data accuracy, and to offer the service to more of their customers.

Company

Aviall, Inc.

Name: Aviall, Inc.

Web Site: http://www.aviall.com

Symbol: AVL (NYSE)

Business:

HQ: Dallas, Texas, U.S.A.

Employees: 817 (2000)

Revenues: $486 million (2000)

Solution

Category: SCM and Warehouse Management System

Application: FreeStock Program

>> Automates the parts inventory and replenishment process for Aviall customers

Technology: CellExchange

>> Allows for wireless integration of the inventory data in the field with the enterprise order system at the Aviall headquarters

Target Audience: Aviall Sales Force

>> 50 customer warehouses in the United States

Devices:

>> Symbol SPT 1700 Palm O/S handheld computer

Challenge and Business Drivers

Business Drivers

>> Desire to provide even higher levels of service to customers
>> Need to automate inventory process and improve data accuracy
>> Need to reduce time in performing parts inventories at customer warehouses

Former Process

Manual counting of inventory for 200 to 300 parts per customer

Benefits

Improved Productivity

>> Automated inventory and replenishment process
>> Reduced time required to perform parts inventory

Reduced Costs

>> Reduced administrative expenses

Increased Customer Service

>> Improved data accuracy
>> Service rolled out to more customers

Challenge

Aviall is transforming its business from an aviation parts distributor into a next generation supply chain services provider to the aviation industry. Part of its business strategy is a consistent focus on its customers. According to Joe Lacik, CIO, one of the key focuses for any new initiatives is to always ask whether the new service will deliver improved value to the customer.

The FreeStock program provides value for Aviall customers by providing a free service for monthly inspections of part quantities at customer warehouses. Aviall currently provides this service to approximately 50 of their high-volume customers and has many of their 80 sales representatives performing the monthly inspections. The original process involved a sales representative going to the customer warehouse and manually counting part quantities for 200 to 300 distinct parts per customer. This process was extremely labor intensive and often took half a day to complete. Demand for the service from other customers was strong, but Aviall found it difficult to extend the service to more customers owing to the time intensive nature of the process.

Aviall wanted a way to increase their level of service for their existing customers using the FreeStock program, to improve data accuracy while performing inventories, and to shorten the cycle time of the process in order to gain sufficient bandwidth to deliver the service to additional customers.

Solution

Aviall knew they needed an automated solution for the FreeStock program. They implemented a wireless solution using software from CellExchange together with Symbol SPT 1700 handheld devices in order to better serve their customers and automate the process. Their approach was to ensure strong user acceptance by first working with CellExchange in order to develop prototypes that could be demonstrated to end users. This interactive approach helped them to refine the functionality on the device and test the solution early on. Since the Symbol devices were equipped with bar code scanners, Aviall was able to place bar codes containing part information on each part bin at the customer warehouse and use this to streamline the inventory process. The Symbol devices also contained the minimum and maximum part quantity information so the sales representatives could easily determine whether a part was under-stocked or not.

While the Symbol devices already contained a wireless modem, they decided to leverage both wired and wireless connectivity since in some customer warehouses it was difficult to gain a strong wireless signal. The Symbol devices were able to capture the inventory information and then integrate with the Aviall back-end replenishment and enterprise resource planning systems.

Benefits

Aviall estimates cost savings of over $1M due to the reduced time required to perform these parts inventories at customer warehouses. They have seen the time requirement drop from about half a day to under one hour. Additionally, the solution has enabled them to increase customer satisfaction with more accurate reporting and more timely restocking, and to offer the service to more customers. Aviall sees this project as phase I of their plans for the warehouse management and inventory control services that they offer to their customers.

One of the lessons learned during the exercise was that Aviall wanted to host their wireless middleware internally. They had the option of hosting via an ASP model, but decided to in-source the solution owing to security concerns. Hosting externally would have meant they would have to open up a path for the wireless middleware ASP platform to access their enterprise applications through their corporate firewall. While the overall solution had strong user authentication and access control on the end user devices, they wanted to keep their own access control in place over the server-side of the solution.

Creating a Wireless Proof-of-Delivery System for Employees and Customers

Office Depot
http://www.officedepot.com

Executive Summary

Office Depot, founded in 1986, is the world's largest seller of office products. They have 961 stores in 18 countries and deliver more than 100,000 shipments to customers across the world every day. Within the U.S., the company has a private fleet of over 2,000 trucks that deliver to consumer and business customers through orders generated from its 829 North America retail stores, a state-of-the-art public Internet site as well as a customized Internet site for corporate customers, and more than 25 million print catalogs are mailed annually in the U.S.

With a vision of being the most compelling place to work, shop and invest, the company has streamlined this vision through its distribution network in order to achieve this mission. More than two years ago, the company embarked on an M-Business customer service initiative to automate and simplify their delivery process and enhance customer service.

The company needed an automated delivery management system in order to provide accurate delivery status information to customers, to eliminate proof-of-delivery write-offs, and to reduce administrative expenses. The Office Depot Signature Tracking and Reporting System, OD S.T.A.R., was developed and implemented in conjunction with Aether Systems and Symbol Technologies by using its SPT 1700 handheld computer. The wireless unit is used in order to track shipments throughout the delivery process. In addition to increasing customer satisfaction, the company realized a return on investment within three to four months at each stage of deployment, and rolled out the solution across the entire company in the U.S.

Company

Office Depot, Inc.

Name: Office Depot, Inc.

Web Site:
http://www.officedepot.com

Symbol: ODP (NYSE)

Business: Office Products

HQ: Delray Beach, FL

Employees: 48,000 (2000)

Revenues: $11.57 billion (2000)

Solution

Category: FFA and Delivery Management System

Application: Office Depot Signature Tracking and Reporting System (OD S.T.A.R.)

>> Automates the package delivery process for Office Depot and customers to better manage and track their shipments

Technology: Aether

>> Information captured in the field is routed through Aether's network operations center to the Office Depot Web site

Target Audience: Employees and Customers

>> Fleet of more than 2000 trucks serving 795 superstores in the United States

Devices:

>> Symbol SPT 1700 Palm O/S handheld computer

Challenge and Business Drivers

Business Drivers

>> Need to provide accurate delivery status information to customers

>> Need to eliminate proof-of-delivery write-offs

>> Need to reduce administrative expenses for managing the delivery process

Former Process

Manual, paper-based tracking systems for orders, invoices, and deliveries

Benefits

Improved Productivity

>> Automated delivery process
>> Increased driver productivity

Reduced Costs

>> Reduced driver overtime requirements
>> Reduced administrative expenses
>> Eliminated proof-of-delivery write-offs

Increased Customer Service

>> Real-time access and visibility into the delivery process

Challenge

Office Depot needed an automated delivery management system in order to provide accurate delivery status information to customers, to eliminate proof-of-delivery write-offs, and to reduce administrative expenses. The former process involved manual, paper-based tracking systems for orders, invoices, and deliveries. With the paper-based delivery tracking system, customers calling Office Depot for order status information were limited to knowing that their deliveries would arrive sometime during the day, but could not gain more precise information about delivery timing.

Solution

After evaluating several wireless solutions, it became clear that the solution to the challenge was the implementation of Office Depot Signature Tracking and Reporting System, known as OD S.T.A.R. Developed and implemented in conjunction with Aether Systems and through the use of Symbol SPT 1700 handheld computers, the company was able to track shipments throughout the delivery process.

According to Dennis Andruskiewicz, Senior Vice President of Distribution, drivers begin their day with a full route manifest loaded onto their devices. After each delivery, they can scan the delivered packages and capture customer signatures electronically. Before their next delivery, they place their device into a docking station and have the truck transmit the delivery information over the wireless network to the office. The majority of delivery information is sent over the wireless network with the exception of signatures. At the end of the day, drivers download this information due to the large file size and less critical importance from a customer service perspective.

At the end of the day drivers can input commands on their devices to have their information uploaded over the wireless LAN network to the company's back office systems. Drivers then print and pick up a reconciliation sheet, which summarizes the day's deliveries and submit along with their checks to the reconciliation office.

According to Kevin Conklin, Project Manager for the OD S.T.A.R. initiative, the solution has been rolled out to all 20 distribution centers, select satellite locations, and approximately 2,100 drivers. Office Depot was an early adopter of onboard technology for their drivers and spent considerable time ensuring a highly successful deployment.

The company started a year and a half ago by looking at the variety of devices on the market. They decided upon the Symbol SPT 1700 owing to its compact size, robust casing, and Palm operating system. They spent approximately three months working to develop the specifications for the delivery management application and another three months in interactive development. The next step was to demonstrate the application for their most important audience—their drivers. They gathered fifty to sixty drivers in order to review the application and to gauge the realistic usability and feedback. The team wanted to ensure that the application was driver friendly and to provide a forum for answering questions on how to use the device and plan for next steps in deployment. The company spent a great deal of time with the drivers on their routes in order to understand the full process in the former paper-based world. After eight months of research and evaluation with drivers, they deployed the application to five drivers during a beta pilot. Following the pilot, several weeks were spent with these five drivers to see how the process was working. This included conducting time studies that focused on time required for downloads and other device-related activities. After this diligent initial beta test, Office Depot rolled out the application in production to groups of ten drivers weekly. This technique allowed them to spend time with each group of drivers and help them use the new system.

The solution is now deployed across the company's entire fleet and has been strongly received by the drivers. According to Andruskiewicz, the drivers were involved in many aspects of deployment. When we were first planning our implementation, we held a contest for our employees to name the application. After receiving hundreds of suggestions, Office Depot decided upon OD S.T.A.R.

Benefits

Benefits realized from using the new solution included improved productivity, reduced costs, and increased customer service. For the drivers, Office Depot was able to eliminate their stack of paperwork containing delivery information. Typically, this consisted of a half-inch thick stack of paperwork per driver per day. Now, the only paperwork drivers carry is a small packet of materials as a backup. The productivity benefits included a time savings of 30 to 50% usually spent by drivers when loading deliveries at the beginning of each day. Also, an administrative time savings of 60 to 70% at the end of the day previously spent in researching paperwork and in performing delivery reconciliation.

Other benefits include the ability for drivers to know their total number of stops for the day (the application counts down the number of stops as they go) and managers can be more pro-active about sending relief to drivers who are overloaded with deliveries on a particular day. In addition to the main emphasis on customer satisfaction, Office Depot has seen a return on investment of between three and four months for their solution at each deployment. Since the rollout of the OD S.T.A.R. application for the drivers, Office Depot took it one step further by integrating direct access to order status information for their customers via their Web site at www.officedepot.com. Now customers have the same access to delivery data as the customer service representatives. It has been a real home run with our customers, concluded Andruskiewicz.

Office Depot's careful rollout of the OD S.T.A.R. application is a textbook example of how to ensure user acceptance, target the precise functionality required through strong user involvement, and generate the most benefit from the overall solution.

Improving Sales Force Productivity Via Mobile Access to Customer and Product Information

Rental Service Corporation
http://www.rsc.com

Executive Summary

Rental Service Corporation is a company within the rental service business area of Atlas Copco. Altas Copco is an international group of industrial companies headquartered in Stockholm, Sweden. RSC provides rental equipment to the commercial construction industry and has a total of 700 sales representatives in 42 states within the United States. The company wanted a mobile application that could help them increase sales force productivity and customer service. The former process involved data synchronization via a toll-free access number and cost the company a considerable amount of time and money for annual connection fees. They implemented an advanced mobile solution for 450 of their sales representatives using mobile software from @Hand and NEC MobilePro 800 handheld PCs equipped with modems. The application reduced connection times by 96%, increased revenues by $12,000 per month per sales representative, improved data validation latency from 48 hours to 1 hour, and achieved an ROI within 4 months.

Company

RCS

Name: Rental Service Corporation

Web Site:
http://www.rentalservice.com

Symbol:

Business: Construction and Heavy Equipment Rental

HQ: Stockholm, Sweden (Atlas Copco Group), Scottsdale, Arizona (RSC)

Employees: 27,000 (Atlas Copco Group)

Revenues: $5 billion in 2000 (Atlas Copco Group)

Solution

Category: SFA / CRM

Application: Mobile access to customer and product information

>> 500 contacts for sales rep to interact with on a monthly basis
>> Access to customer, product, territory, new lead, and order status information

Technology: @Hand

>> Integration with AS/400
>> Server-side business logic on SQL Server
>> Support for online and offline usage

Target Audience: Sales Representatives

>> Total of 700 reps in 42 states
>> Solution deployed to half of sales team

Devices:

>> NEC MobilePro 800 handheld PCs with modem
>> Windows CE operating system

Challenge and Business Drivers

Business Drivers

>> Sales force information needs and productivity

>> Efficient sales process

>> Customer service

Former Process

Data synchronization via toll free number access

Benefits

Improved Productivity and Time Management
>> 2 hours per day

Increased Revenues
>> 10% increase in monthly revenues generated from the field
>> $50 million projected in the first year

Reduced Costs
>> Connection times reduced by 96% when performing data transmission

Faster Cycle Times
>> Data validation latency improved from 48 hours to 1 hour

ROI
>> Less than 4 months

Challenge

Rental Service Corporation is a company within the rental service business area of Atlas Copco. Altas Copco is an international group of industrial companies headquartered in Stockholm, Sweden. RSC provides rental equipment to the commercial construction industry and has a total of 700 sales representatives in 42 states within the United States and is headquartered in Scottsdale, Arizona. The company wanted a mobile application that could help them increase sales force productivity and customer service. The former process involved data synchronization via a toll-free access number and cost the company a considerable amount of time and money for annual connection fees.

According to Jeff Cummings, VP of Strategic Accounts at RSC, the project arose as a result of work conducted by the sales best-practice steering committee. The committee had decided that a sales force automation program was in order and had submitted the request to their executive committee. While return on investment was asked by some of the upper management, the initiative was funded based upon the need to better manage information for the sales force, to improve productivity, and to increase customer satisfaction. Jeff concentrated on being an ambassador for the solution and evangelized it from the user perspective within RSC despite some challenges from his IT department regarding the feasibility of wireless and mobile solutions.

Solution

They implemented a custom sales force automation solution for 450 sales representatives using mobile software from @Hand and NEC MobilePro 800 handheld PCs equipped with modems. The application provided access to customer, product, territory, new lead, and order status information. Initially, RSC had looked at the MobilePro 750 model and had given them out to about 40 district sales managers. The sales managers ended up choosing the 800 model owing to its larger size despite otherwise similar features.

RSC wanted an application that could work in both connected and disconnected modes. The RSC sales force worked in both urban and rural areas and so could not depend on gaining wireless connectivity all of the time owing to coverage issues in rural areas. The @Hand technology provided them with the means to capture data while out of the coverage area and then synchronize the data with the back office when the connection was restored. The back office data source was an AS/400. The handheld PCs were chosen for their larger screen size, full-sized keyboard, and longer battery life when compared with the smaller PDAs such as Palm Pilots or Pocket PCs.

According to Jeff, as the solution was put into beta test, there was a lot of interest from sales management in adding extra functionality to the application. One way he kept the scope of the project tight and validated new functionality was to keep going back to the question of whether the functionality truly benefited the point of sale.

He wanted to place report capabilities in the hands of the sales staff so they could have access to the same information as their management. The application was also designed in order to pro-actively walk the sales staff through the sales process in the various construction phases of their customers.

The solution was rolled out with a built-in modem in the device that uses a phone line for synchronization. Some sales representatives also have wireless connections—either hard-wired in their vehicles or via cellular modem. The majority, at the present time, however, use a standard phone line and synchronize at the end of the day. One of the reasons for this when compared to a more real-time wireless connection was that much of the data had a refresh rate of once per day. Synchronizing once in the evening allowed the sales staff to upload and download the relevant information (just the changes) without missing anything on a more granular frequency. For example, leads typically came in once per day. If the salesperson synchronized once in the morning, they would have those leads on their device the same day.

The phased rollout of the solution included an initial beta test of the application with five to six people on the best practice committee, a second beta test with twenty sales people in the field, an introduction and training session on the devices and resident software such as e-mail and word processing software at a national sales meeting, and finally a district-by-district training session focused on the sales force automation software itself.

Benefits

The estimated productivity increase for the sales representatives was up to 2 hours per day or 25%. The productivity increase was realized owing to the automation of their field transactions and processes to eliminate paperwork and administrative tasks such as reducing the time spent dialing in to a toll-free number to perform data transmission. In fact, annual connection times for the entire sales force were reduced from over 3 million minutes to under 250,000 minutes when the @Hand solution was implemented. Prior to the @Hand solution, each sales representative would spend on average 20 minutes per day in connection time in order to synchronize their devices. After implementation of the @Hand solution, the connection times dropped to around 2 minutes per day. The main reason for this was that the @Hand software transmitted just the changes to the data and not the entire datasets between the server and the client devices. This was a straightforward feature, but one that yielded significant time savings for the sales force. The @Hand software also provided RSC with improved centralized management of their application.

Overall, the mobile application reduced connection times by 96%, increased revenues by $12,000 per month per sales representative, improved data validation latency from 48 hours to 1 hour, and achieved a return on investment within 4 months. The additional $12,000 per month per sales representative translated into an estimated $50M per year in increased revenues for RSC.

RSC was able to measure the increase in productivity of their sales force by comparing those with the handheld devices with those not using the devices. The other variables such as the geographic market area and the rental equipment offered were kept consistent across the mobile-enabled and non-mobile groups by looking at teams from the same stores or branches. When looking at the total revenue per month for each salesperson, Jeff found that the mobile-enabled sales staff were generating 9% to 11% more revenue than those without the application.

One of the unexpected benefits of the solution turned out to be sales management and reporting. Since the application allowed the sales force to better track and manage their territories, RSC sales management was able to better allocate these territories among their sales staff. Thus, there were less variances between individual sales personnel.

The application has also provided competitive advantage through faster lead distribution. RSC gets some of their leads via a third party who provides them with approximately 800 leads per day. RSC downloads these leads to the user's device as part of the nightly synchronization operation. One of their competitors has the leads mailed to them, which can take up to a week to receive.

In summary, the key benefits to the sales force have been the ability to better manage their time, their territories, and their leads. With each salesperson typically having 120 job sites and 5-10 contacts per site, there is no way to manage this information efficiently on paper and the application has given them a powerful management tool.

Public Services

In the public services area, M-Business can be applied to vast numbers of application scenarios within local-, state-, and federal-government, within the military, and within the various public agencies such as fire departments and law enforcement. In many cases, access to information can be a matter of life and death, as well as providing improved efficiency of operations and community relations benefits. Law enforcement officers and rescue crews are able to better understand and react to critical situations if they have an improved ability to communicate with one another and with their headquarters. Mobile access to various information databases can provide them with critical information about potentially dangerous situations they are entering into.

For example, a rescue crew could receive data on the layout of a burning building. This could help them navigate through the building to rescue trapped victims. A law enforcement officer could pull up information using a wireless handheld device related to a person, vehicle, or other article from a major criminal database without having to make trips back to the officer's vehicle. A fire department could perform property inspections for potential fire hazards and have inspection data immediately validated upon entry on a wireless device.

Health Care

Creating a Mobile Solution for Doctors

University Hospital (Universitaetsklinik), Mainz, Germany

Executive Summary

The neurosurgery department within the University Hospital in Mainz, Germany, was looking for a way to increase doctors' productivity by allowing them to spend more time focused on patient care and less time manually looking up and writing down classification codes to satisfy German regulations.

They used a Bluetooth network to enable flexibility and mobility around the hospital so that doctors could manage patient records via their PDAs. The solution constantly synchronized data wirelessly and gave the doctors access to the most recent patient information.

Company

University, Mainz

Name: University Hospital, Mainz

Web Site: http://www.klinik.uni-mainz.de

Department: Neurosurgery Department

Business: Healthcare

HQ: Mainz, Germany

Solution

Category: Wireless LAN

Application: Wireless access to patient records

Technology: Red-M Bluetooth Network

Target Audience: Physicians

Devices: PDAs

Challenge and Business Drivers	Benefits
Business Drivers	Time Savings
>> Productivity	Greater Data Accuracy
>> Data Accuracy	More Time On Patient Care
Former Process	
>> Physicians had to manually look up codes and make paper-based notes	

Challenge

In order to satisfy German regulations, all doctors at the University Hospital have to record diagnoses and surgical procedures for patients according to international classification codes stored in the hospital's database. Doctors had to take time away from patient care in order to manually look up codes and record them on paper. Later, these paper notes were entered into the system for balancing against the patient's health insurance policies. The Hospital was looking for a way to increase productivity and allow the doctors to have easy access to patent data as they moved around the hospital.

Solution

The solution that was implemented was the Bluetooth networking solution from Red-M, a U.K.-based company that develops and markets hardware and software product solutions that enable mobile data, voice, and video communications inside buildings and public concourses to a range of handheld and mobile devices.

The solution utilized the Red-M 3000AS access server and the 1000AP access point as part of the Bluetooth network. The access points are small devices that are situated around the hospital areas where the Bluetooth network is to be provided. Access points act as "Bluetooth base stations" and provide up to 31,420 m² of wireless coverage around each access point. The access server is the central device in the Bluetooth network and provides shared network services such as Internet access, intranet access, and connectivity to a PBX-based voice system.

Benefits

Some of the benefits of the solution have included the ability for doctors to have access to the most recent patient information on their PDAs, the ability to access classification codes for patient care on their PDAs from data stored in the hospital's SAP R/3 database, and the general increase in productivity that the PDAs have enabled. Doctors can now spend more time on patient care and less time on the less-value added tasks of record keeping.

Financial Services

No other private sector industry relies on the immediacy and accuracy of data and transactions as much as the financial services industry. Because of this, the industry has been one of the first to adopt M-Business in both the consumer space and the business-to-business space.

General examples of M-Business within this industry include the following: consumer access to place trades, retrieve account information and access real-time market data via discount brokerages; access to financial portals providing market news, analysis, and commentary; access to bank account balances with the ability to perform transfers, to pay bills, and to contact customer support; access to hotel availability and reservations; access to insurance agent locations by zip code lookup, claims, and policy service; and billing and payment information.

On the business-to-business side, examples include access to critical notifications regarding time-sensitive financial market data, as in the FT Interactive Data case study, and access to key performance indicators in the hospitality industry, as in the Carlson Hospitality case study.

The following case studies aim to illustrate some of the wide range of solutions that have already been implemented in various sectors of the industry.

Enhancing Customer Service With Time-Sensitive, Business-Critical Customer Notifications to Wired and Wireless Devices

FT Interactive Data
www.ftinteractivedata.com

Executive Summary

FT Interactive Data supplies global securities data and fixed-income portfolio analytical software to the international investment community.

The company needed a solution to distribute time-sensitive financial market data such as late pricing corrections from a stock exchange or product status bulletins more efficiently and accurately.

The prior process relied on a combination of a custom e-mail system and out-sourced fax services to communicate with clients and was time-consuming, expensive, and error-phone. Additionally, there was no way to send out automated voice-based messages.

By implementing an enterprise application messaging solution from EnvoyWorld-Wide, the company increased accuracy in the distribution process, improved customer service, gained a complete audit trail for message delivery, and achieved a 237% return on investment in customer service time and resource.

The solution supported multiple communication means including pager, e-mail, fax, and voice based upon each client's preference.

Company

FT Interactive Data

Name: FT Interactive Data

Web Site:
http://www.ftinteractivedata.com

Symbol: Major operating division of Interactive Data Corporation (NASDAQ: IDCO)

Business: Financial Services

HQ: Bedford, Massachusetts

Employees: 1,600 worldwide

Solution

Category: CRM

Application: Wireless delivery of time-sensitive financial market data to clients

>> Pricing Corrections
>> Product Status Bulletins

Technology: EnvoyWorldWide hosted service

Target Audience: Customers

>> 27,000 messages per month
>> 10 customer service
    representatives

Devices:

>> Pager, E-Mail, Fax, Voice

Challenge and Business Drivers

Business Drivers

>> Improved Customer Service
>> Data Accuracy
>> Increased Productivity

Former Process

>> Customer service representatives had to use a variety of disparate systems such as custom e-mail and outsourced fax services in order to provide clients with notifications

Benefits

Improved Customer Service

Improved Data Accuracy

>> Single system

Increased Productivity

>> Customer Service Reps—4 hours per week
>> Total of 10 CSR's

Cost Savings

>> $6000 per year in mailings

ROI

>> 237% in customer service time and resources

Challenge

A major operating division of Interactive Data Corporation, FT Interactive Data supplies global securities pricing, dividend, corporate action, and descriptive information that supports fund pricing, securities operations, research, and portfolio management. The company supplies data from over 3.5 million securities traded around the world to banks, brokerage firms, insurance companies, money managers, and mutual fund companies.

According to Susan Burud, Director of Customer Service, FT Interactive Data needed a one-stop shop solution to quickly get out pro-active notifications to their clients. The prior process used a disparate collection of outbound systems and distribution lists. Additionally, voice-based messages were needed for the cases when customers were away from their desks and hence away from their e-mail and fax machines. The company needed a solution that would allow them to store and maintain all their lists in one place and to distribute information using the customer's preferences for communication method.

Solution

The solution implemented was EnvoyWorldWide's EnvoyXpress. Using this solution, FT Interactive Data was able to distribute and track the messages sent simultaneously to customers via fax, e-mail, and pager according to the recipient's preferred communications device. The ability for tracking message delivery was another valuable feature, since previously the company had no way to track if the messages were actually received.

Benefits

The return on investment for FT Interactive Data was determined both up-front prior to implementation and also post-implementation. The up-front calculations looked at cost savings, while the post-implementation measurements looked at both cost savings and customer satisfaction.

The ROI formula for FT Interactive Data was as follows:

$$ROI = \text{Total benefits (cost savings) / Total costs}$$
$$= \text{(Labor cost savings + Other cost savings) / Total costs}$$

where

Labor cost savings = (# of reps) x (salary + benefits) x (% time savings) x (productivity rating)

Based on a 40-hour week, the solution saved the customer service representatives about 4 hours per representative/per week, which amounts to a 10% time savings. A productivity rating of 65% was applied, since not all of the time savings translates into additional work. Other cost savings included an estimated $6000 per year saved in mailings.

The EnvoyWorldWide service was charged on a monthly basis, so this figure was used to derive the total cost of the service per year.

The final return on investment from the calculation, with ten customer service representatives using the solution and sending out as many as 27,000 messages per month, came to 237%.

The main benefits for FT Interactive Data included the increased productivity of the customer service representatives, the improved data accuracy, and the enhanced customer satisfaction. According to Susan Burud, one of the new challenges in the solution has been the ability to keep contacts at the various companies up to date. When clients leave an organization, they need to be informed so that they can add a new contact to the system.

Improving Productivity and Customer Satisfaction With Business Intelligence

Carlson Hospitality Worldwide
http://www.carlson.com

Executive Summary

Carlson Hospitality Worldwide wanted to provide managers with more timely and more focused access to critical business information related to their hotel properties. They deployed a wireless business intelligence solution called Mobile Access to Carlson Hospitality, version 1 (MACH-1) in order to provide regional and general managers of their hotel chains with alerts based on key business indicators such as occupancy rates, sales trends, and supply levels. The solution was implemented using Compaq iPAQ handheld devices running the Microsoft Pocket PC operating system. Benefits included faster access to vital information and improved decision-making leading to enhanced property financial performance. Additionally, managers were able to interact more with customers by becoming less desk-bound.

Company

Carlson Companies

Name: Carlson Companies—Carlson Hospitality Worldwide

Web Site: http://www.carlson.com

Symbol: One of the largest privately held corporations in the United States

Business: Travel (Carlson Wagonlit Travel), Marketing, and Hospitality Services (Brands include Regent International Hotels, Radisson Hotels & Resorts, Country Inns & Suites By Carlson, Park Inn and Park Plaza Hotels, T.G.I. Friday's)

HQ: Minneapolis, Minnesota

Employees: 188,000 worldwide

Solution

Category: Business Intelligence

Application: Mobile Access to Carlson Hospitality, version 1 (MACH-1)— Wireless delivery of time-sensitive alerts and notifications

>> Occupancy Rates
>> Average Room Rates
>> VIP Guest Check-In
>> Service Levels
>> Yield Management Controls

>> Revenue Per Available Room (revPAR)

Technology: Custom built solution

Target Audience: Employees

>> Regional Managers

>> Hotel Management

Devices:

>> Compaq iPAQ PDA
>> Pocket PC Operating System

Challenge and Business Drivers

Business Drivers

>> Improved Customer Service
>> Mobile Decision Support
>> Increased Productivity

Former Process

Desk-bound PC systems and paper-based reports such as monthly and weekly management reports

Benefits

Improved Customer Service

Timely and Improved Business Decisions

Increased Productivity

Cost Savings

Alignment With Company Strategy

Challenge

Carlson Hospitality Worldwide is part of Carlson Companies, Inc. based in Minneapolis, Minnesota. The company is one of the largest privately held corporations in the United States and is focused on the travel, marketing, and hospitality markets with brands including Regent International Hotels, Radisson Hotels & Resorts, Country Inns & Suites By Carlson, Park Inn and Park Plaza Hotels, and T.G.I. Friday's.

Carlson Hospitality Worldwide had invested in a three-year $21 million program in order to enhance its core reservations and information-management systems. As a part of this program, they decided to extend the applications and information into the wireless arena for improved employee data access and decision-making ability.

The story behind how this initiative came to life is an interesting one. According to Scott Heintzeman, CIO, Carlson Hospitality, the project started as a combination of two separate business drivers occurring at the same time. Firstly, corporate staff were becoming interested in handheld PDAs and Heintzeman had noticed several purchase orders coming through for PDAs for calendar and schedule use. Heintzeman wanted to see if there was some way to make these devices more strategic beyond simple calendaring and scheduling. At the same time, in a totally separate line of conversation, there was a request from user constituencies to boil down the mass of information that they had to digest into a more usable format and to make the information more pro-active.

Heintzeman and his IT team connected both of these conversations and identified a better, more highly leveraged solution. Two seemingly disconnected situations: The increase in PDA purchases and the need for efficiency in information delivery lead to the idea of taking enterprise information and boiling it down to the critical business indicators and placing it on the handheld devices. Additionally, the information could be delivered more proactively to the end users via user-controlled alerts and triggers on data points in order to push out notifications to the devices.

Heintzeman and his team realized the concept could be applied quite broadly across their organization. The third driver behind the birth of the solution was that the handheld wireless environment was becoming more affordable and possible. The technology was catching up with the two business issues that were driving the initiative.

Solution

The solution implemented was part of a one-year $1 million program named Mobile Access to Carlson Hospitality, version 1 (MACH-1). The solution was custom developed and included Compaq and Microsoft technology. The application was a wireless business intelligence application designed to provide regional and general managers with access to key business metrics and alerts to key business events. The application worked by allowing users to go to their desktop machines and subscribe to key business indicators and establish their performance thresholds. They choose which indicators they want to have set in alert mode and which to have a goal or threshold on. The mobile device displays different colors as indicators move in a positive or negative direction from the threshold setting. Alerts are sent via e-mail pushed out from the application. Users are able to click on a link within the e-mail in order to jump into the data in question.

Typical events monitored include revenue and booking trends, VIP guest check-in, guest service levels, and occupancy rates. The application provides alerts when key metrics fall outside pre-defined levels and also provides graphical views of data. Events monitored included the following:

>> **Revenue Per Available Room (revPAR)**—This is one of the most important numbers for a hotel to manage and is the best indicator of the combination of occupancy percentage and the average rate per room.

>> **Average Room Rate**—The average room rate per hotel or group of hotels.

>> **Occupancy Percentages**—The occupancy percentages per hotel or group of hotels.

>> **Denials**—This indicator measures potential reservations that a hotel turned away owing to high demand. It is an indicator of market pressure or demand.

>> **Accounts Receivable**—This indicator helps Carlson to ensure hotels are paying their bills in a timely manner.

Additionally, all hotels are monitored in their local currency. Japanese hotels in yen, U.K. hotels in pounds, and U.S. hotels in dollars. The rollup analysis for company wide reports is reported in U.S. dollars.

The devices used by Carlson were Compaq iPAQs running the Microsoft Pocket PC operating system. The MACH-1 application incorporates information from several Carlson back-end systems including their HARMONY property management system, the CustomerKARE system, the Curtis-C Reservation System, and KnowledgeNet. Additionally, the MACH-1 devices also provide access to e-mail, Internet access, the ability to send a voice mail by using the iPAQ's built-in microphone, and the ability to exchange information (contacts, tasks, calendar events, notes, and files) with Palm devices.

The rollout of the solution consisted of three phases: the IT department, the Carlson corporate users, and then to hotel operators and owners. The IT team was leveraged for a practice run in order to confirm the reliability and usability of the solution with about 20 staff members. The solution was then deployed to the main line of corporate staff. About 150 devices were distributed to staff in the Carlson corporate office, which included regional and general managers for the Radisson Hotel chain. The next wave was testing in a handful of hotels where Carlson piloted the device in live hotel operations. Eventually, Carlson plans to more broadly deploy the solution to partners in corporate offices and hotel installations including the other Carlson hotel brands such as the Regent Hotels.

Carlson has filed a patent on five components on their solution: 1) use of wireless technology; 2) monitoring of dynamic business information; 3) where that information is subscription-based to a user's personal preferences; 4) ability to set thresholds on any of the information to which they subscribe; and 5) where they send alerts and trigger alarms when information exceeds a threshold. They have identified over twenty applications of these five core characteristics, including applications for monitoring financial information, operations, and customer satisfaction.

Benefits

Benefits from the MACH-1 implementation have included faster access to vital information and improved decision-making, which has lead to enhanced property financial performance. Additionally, managers are able to interact more with customers by becoming less desk-bound to their PC systems and monthly and weekly reports. Ultimately, Carlson expects this implementation to translate into better customer service for all guests and enhanced financial performance for all of their hotels.

Lessons Learned

According to Heintzeman, some of the lessons learned from the MACH-1 project have been as follows:

>> **New Way of Thinking Was Required**—Initially the invention was thought of as a novelty. It was important to help users see the solution as a new way of working and as a serious business tool. This included changing the way they wanted to look at their business information. Instead of expecting to look at multi-row and multi-column spreadsheets, the users were asked to determine what was really important to them in terms of key fields. This was done so that this information could be boiled down into mini-reports and the five to ten indicators they really cared about could be implemented as triggers and alerts.

>> **Devices Became a Portable "Balanced Scorecard"**—The MACH-1 devices were found to become the portable balanced scorecard within the company—providing a wireless, portable, real-time, customized way for individuals to monitor and achieve their personal goals for each day.

>> **Ability to Monitor Key Indicators Provided Business Agility**—The MACH-1 System gives new meaning to the concept of "managing by walking around." Each manager has the ability to subscribe to only the information that they need to do their job, thereby enabling people to be mobile, focused, informed, and highly effective "knowledge workers."

>> **Real-Time Nature of the Devices Improved Enterprise Data Quality**— When Carlson started watching the data coming out in real-time and being drawn from major enterprise systems, they started to see certain data management processes that were broken and which could not handle the rigor of the near real-time reporting. They had to re-tool the way that data was captured and redefine the meaning that was attached to the data. The MACH-1 application became a poster child for doing things right within the IT department.

>> **Wireless Bandwidth was Still Problematic but did not Slow Down the Solution**—Carlson provided three ways for users to synchronize their data on their devices: via the cradle or IR port, via an 802.11 wireless LAN, or via a

cellular call. The most popular solution for end users was the cradle synchronization. Employees in the corporate office in Minneapolis often leave their devices on sync within the cradle and then take the devices with them to meetings. Carlson is exploring all connection options and aims to migrate away from the sync solution into more real-time connections as communications technology evolves. What has been important has been to get the technical platform solid and to drive the behavior change while letting the bandwidth issues catch up. Carlson felt it was important to learn and practice early on in order to stay ahead of the competition.

>> **The Application Renewed Investor Interest**—Carlson has noticed tremendous interest from owners and franchisees in MACH-1. Investors and operators are increasingly more sophisticated. They have studied what technology can do in the hotel industry and are aware of the many systems for hotel property management, reservation services, customer information systems, electronic campaign management, and so forth. However, when these owners and investors saw the MACH-1 solution, it generated very positive feedback and renewed excitement about investing in Carlson's hotel brands.

>> **A Handheld Backup for Property Management Systems**—In one of the hotels, using an application that provided information from their property management system, the staff told Carlson that they discovered a new interesting feature of the application. When their property management system had to be taken offline suddenly for maintenance, the handheld devices became an unexpected hot backup of their system's crucial information on guests and house status. Because MACH-1 was perpetually refreshing with the property, the users had a portable snapshot of their property management system from which to work until the core system came back online.

6

M-Business Strategic Roadmap

An M-Business strategy can help the enterprise increase its business agility by applying the right M-Business applications and processes to the right areas of need and opportunity across all user constituencies. The strategic approach can help to maximize enterprise benefits, to minimize the risks inherent in any technology-related initiative, and to ensure executive buy-in. Another advantage of an M-Business strategy is that M-Business initiatives can be aligned with broader business strategies and can be prioritized based upon their potential for providing the maximum benefit to the enterprise.

There are also benefits in approaching M-Business strategy from a corporate-wide viewpoint, as opposed to a departmental viewpoint. Implementing a corporate-wide M-Business strategy can yield considerable advantages over rolling out M-Business applications on a case-by-case basis. It presents the opportunity to leverage the synergies between applications from both a technical and business perspective. This creates an overall value that is greater than the sum of the individual parts.

An M-Business strategy should take into account both the "as-is" environment and the "to-be" environment. It should leverage existing investments in applications and skill sets, but also provide a clear path toward the "to-be" business objectives.

M-Business projects may be completely new business initiatives or they may be extensions of other IT projects that have both wired and wireless aspects to the overall solution.

Thus, some of the goals of an M-Business strategy should be as follows:

>> Envision M-Business initiatives that provide true enterprise value for employees, customers, suppliers, and business partners

>> Determine which projects to fund based upon strategic value and return on investment

>> Determine project prioritization based upon business impact and benefits

>> Ensure alignment with Business Strategy

>> Ensure alignment with IT Strategy

>> Maximize existing technology investments and skill sets

>> Leverage of M-Business portfolio across the enterprise

>> Ensure maximum return on investment for M-Business initiatives

>> Maximize the potential for project success by managing risks and determining critical success factors

Since every company has its own approach and methodology for conducting strategy work and for funding project initiatives, the following sections will focus more on recommendations specific to M-Business than in providing a specific methodology to follow. You can leverage these recommendations however it makes sense within your own approach and methodology.

We will group the recommendations into sections on business and technology assessment, identification of target opportunities, prioritization of initiatives, business case development, and project plan development.

Business and Technology Assessment

In performing an M-Business strategy, you will often want to start by exploring and documenting the "as-is" and "to-be" environmental context in which your initiatives need to exist and provide value. This information can often be captured via a combination of interviews, workshops, and review of existing documentation and standards.

As-Is Environment

One of the first initiatives to perform is the assessment of the current state business and technology environment. This can help lay the groundwork so that the current "as-is" environment is well understood and documented. In a large corporation, this can be quite a challenge and it is important to tailor the amount of effort in this phase to the specific level of need. Generally, you will want to capture enough information to have an inventory of current application initiatives across the enterprise and an inventory of current skill sets, application standards, best practices, and guidelines. The technology assessment should give you enough information to understand what is in place today and what can be leveraged in future M-Business initiatives.

With regard to M-Business applications, you may wish to determine the range of wireless and handheld devices in use within the enterprise and any current wireless applications that are in use or in a pilot stage. Determine which devices are employee-owned and which are supported by the enterprise. Also look outside the enterprise to ascertain what devices such as pagers, PDAs, and handhelds are frequently used by your customers, suppliers, and business partners. With this knowledge, you may be able to roll out M-Business applications that support the currently available devices without having to require additional purchases and new training. Having this information at hand provides an excellent frame of reference as we move forward in our M-Business strategy, identify target initiatives, and prioritize opportunities. Note that it is also important to balance the relative merits of using existing devices with the merits of moving toward newer devices with improved functionality. Certain M-Business applications may have limited functionality on the currently available devices, but may have the potential for greatly enhanced functionality on some of the newer devices. This speaks to the age-old issue of building for the lowest-common denominator or building for a more feature-rich newer device or standard. Part of this question may be answered by understanding your end users and determining if and when they are prepared to migrate toward newer devices and standards.

On the business assessment side, it is important to gain an understanding of what applications are currently in use by the various user constituencies, the processes associated with these applications, and the current pain points around these processes. What are the core groups of user constituency within the enterprise? What are the metrics around these user constituencies and how are they measured? What key performance indicators are used to measure effectiveness? Where are these users located geographically and what is their degree of mobility?

Understanding the behaviors and processes of your target users will help to determine the level of enterprise benefit that can be realized and also the amount of effort required in order to effect the change in behavior. Capturing some of the metrics around current application usage can also be useful when determining the return on investment for the new M-Business application. You will be able to compare the old and new process models and estimate the time savings and other productivity factors that may apply. A field visit can be essential and can also serve to uncover new opportunities for business agility that may have been previously missed.

To-Be Environment

The business and technology assessment phase should also take a look at future directions in strategy. Determine the future "to-be" strategy for the business and for the technology side of the house. What are the business objectives this year and next year? Is the company focused on customer satisfaction, cost reduction, productivity improvement, revenue generation, developing new products and services, or a mixture of some or all of these? What are the priorities? What direction is the company taking with regard to information technology? What standards are being adopted and which are being phased out? What new skill sets are being developed? What applications are being insourced versus outsourced? Which application areas has the company deemed as most strategic to the business? What new packaged applications are planned? What emerging technologies are being studied for future enterprise benefit? What is the timeline of adoption for these technologies? What software and services vendors is the company currently working with in order to evaluate new products and services?

When the assessment of the "as-is" and the "to-be" environment is complete, you should have a clear picture of the current and future business and technology landscape within your company—at least to the level of detail required by your strategy. How broad a view you choose to take depends on whether your M-Business strategy is corporate wide, aimed at a specific functional area or user constituency, or is focused more narrowly on just one or two areas. The timeframe for this phase will obviously vary based upon the scope, but in general it should be a fairly rapid phase in the order of a few weeks. Much of the required information will be readily available and already documented. Some of it may need to be determined from interviews and conversations with various stakeholders.

Identification of Target Opportunities

From an understanding of the current and future directions in terms of both business and technology strategy, one can start to look around and identify areas of opportunity and areas of need where M-Business can enhance enterprise value. Some areas of opportunity may be readily apparent. Others may only be uncovered by understanding the overall business and technology landscape within the enterprise and by connecting the dots between seemingly un-related points of need.

As we have seen in the earlier chapters, areas of opportunity may well lie across all your user constituencies. Applications such as customer relationship management, sales force automation, field force automation, supply chain management, and enterprise resource planning may all stand to benefit from increased business agility brought about by M-Business applications. Look for opportunities to remove bottlenecks in information and transaction flow. Which user groups are most mobile? How can they become more productive while mobile? How can data accuracy be improved by collecting information at the point of activity versus capturing it manually and using scarce resources to perform data entry later on? What business processes can be streamlined? Which business processes are most critical to customer satisfaction? Which processes are most critical to enterprise operations? Where are the bottlenecks within the supply chain? How can vendors and suppliers be better managed in terms of quality assurance? What information are sales reps and field service reps asking for while in the field? What information can help them provide better service to their customers? What information are your customers asking for in real-time? Would they benefit from real-time, anywhere access to order entry and order status information? Can you provide new levels of customer service and support by leveraging wireless devices and technologies such as SMS?

These are just a few examples of the types of questions to be asking when identifying target opportunities. Hopefully, the case studies and examples of application areas within the book will give you ideas for the kinds of questions to be asking within your enterprise. In the next section, we'll look at how to distill the opportunities that are identified into a set of the most high-priority initiatives that will have the most benefit for the enterprise.

Prioritization of Initiatives

In the previous phase, you may well have identified numerous opportunities for M-Business initiatives within the enterprise. The next step is to determine which initiatives warrant further development and how they should be prioritized. To prioritize the initiatives, you'll want to look at several factors. These include the following: the return on investment; the degree of fit with current business strategy and objectives; the degree of fit with current IT strategy; the total cost of ownership; the competitive environment; the risk factors involved with the project; the level of pain and need around this specific solution; the market demand if this is an external initiative; and the availability of resources and skill sets to tackle the initiative.

A cost/benefit matrix may also be a useful way to visualize which applications provide the highest benefit with the lowest cost. Those that provide high benefit with minimal investment are obviously quick wins. Those that provide high benefit with larger investment costs and higher risks are typically the more strategic applications that lie on the upper end of the risk/reward continuum. Typically those applications that are low risk provide less reward, but in the M-Business arena you may be able to uncover some exceptions. For example, you may be able to wirelessly enable an existing customer relationship management or sales force automation package with a relatively small amount of effort and generate valuable business benefits. Many of the enterprise case studies we profiled earlier, provided solid benefits by focusing on a small application area that was a quick win for customers and employees.

When prioritizing your application initiatives, you will also want to look at the realistic timeline for deployment. Certain applications may be deployable today given the current state of the technology and market acceptance, whereas others may be further out on the horizon. For example, customer access to order status and customer service information may be a good prelude to eventual customer access to M-Commerce application functionality. For the wireless carrier, customer access to unified messaging functionality may be a precursor to further wireless data applications—such as a wireless data portal, M-Commerce applications, and wireless application service provider offerings.

The initial applications that are targeted should not only be self-justifying in terms of the return on investment, but they should also be considered within this portfolio approach where they help to set the groundwork for future applications to build upon. A good example of this is the executive dashboard and business intelligence applications. These types of applications can provide enterprise access to key business metrics, but can also set the stage as a portal or dashboard for future expansion of application functionality.

Business Case Development

Once you have identified and prioritized your M-Business applications, you may wish to develop a mini-business case around the most compelling initiatives in order to gain project approval and funding. In developing the business case, there will certainly be many differences based upon whether the initiative is internally focused or externally focused and whether this is an extension to an existing offering or an entirely new product or service. At the present time, a business case for an internal initiative is a much easier proposition, since the audience and the level of pain or opportunity is well-established and often well-articulated.

Focusing on consumers right now is perhaps the hardest proposition owing to the uncertainties in the market sizing and the potential adoption rates. But even with a well-defined target audience as in the case of enterprise employees, it is still very important to think about how user acceptance will be achieved and how employees will be trained to use the applications. Employees may face several hurdles here. Some of them may be using handheld devices for the first time, so there is a learning curve on both the devices and the applications.

Companies focusing on consumers or external customers often tend to lack focus when developing their business case for entering new markets or delivering new products and services. It is very important to know exactly what industry is being targeted and what user role within that industry will be using your application. You should determine the ideal industry characteristics and the ideal user profile if you have a service offering and are attempting to determine the best point of entry. This is often true for wireless application service providers who have fairly horizontal solutions that can be applied to a variety of industries. For these companies, it is important to understand the competitive landscape and the degree of fit between various industries and horizontal solutions. For example, alerts and notifications may be business-critical in certain industries such as travel, transportation, security, and financial services (where information has a high level of time dependency), but are less critical in others. But even once you have identified the target verticals, your business case will need to also consider who the end user will be and how they will use the application in their work activities. Making the business case as precise as possible will save time and effort later, since pilots and full-scale implementations will be able to focus the right application at the right user, at the right time with the right functionality.

Some of the major sections include to include are as follows: market sizing; the target industry; customer and user role; the ideal customer profile; the current process; the current problem or opportunity; the value proposition of the solution; the solution description; the competitive environment; the team who will implement and support the solution on an on-going basis; and the financial analysis and the risk factors. The business case should also include recommendations for next steps in the process as well as a proposed schedule. The value proposition for the enterprise should be articulated in terms of both hard and soft benefits. That is, it should explore the potential of solid revenues or cost savings, as well as less tangible but equally important factors such as customer satisfaction, customer loyalty, and differentiation from the competition. The section on return on investment later in this chapter should help to provide some tools for calculating both return on investment and net present value of application initiatives.

Project Plan Development

As part of your M-Business Strategy and business case, you may develop high-level project plans in order to scope out the amount of effort required and the project timelines. Putting together a project plan can help you to think through the resource requirements, major activities, milestones, and deliverables. It will also help you to scope out the amount of functionality to be included in the application and the amount of work required in order to build or implement this functionality. Detailed knowledge of application requirements in terms of hardware, software, network connectivity, and ongoing support costs can help you to make a good estimate in terms of the total cost of ownership of the solution and the potential return on investment. The project plan will help you to add in the resource costs in terms of staff committed to the project. It will also help you to determine the appropriate touch points with end users and stakeholders in order to ensure a smooth rollout of the application.

Depending upon the nature of your implementation, whether it is a pilot or a full-scale production deployment, your project plan will have some or all of the following major phases: requirements analysis; architecture and infrastructure planning; design, development, or package implementation and customization; testing; and deployment. In the following two chapters, we'll look at some of the technical architectures for M-Business applications and some guidelines for implementation.

Return on Investment

Return on investment is a term that seems to come and go in cycles. It is always important, but is often given more prominence, especially in the media, when the economy is slower. The true fact of the matter is that return on investment and its importance varies from one corporation to the next. Even in the height of the Internet and E-Business boom cycle, some more conservative companies would still require a strong ROI to be demonstrated prior to funding a project. Others would feel the need to invest purely for the objective of keeping up with the competition and the "new entrants" that potentially threatened their markets. The last several years, where return on investment was somewhat sidelined, can be considered a luxury. Internal project sponsors, software vendors, and systems integrators had an easy time selling solutions into the enterprise during this time.

With the threat of new entrants diminished for the old world companies after the bursting of the dot.com bubble, companies went back to ROI as a key factor when looking at most if not all projects in addition to their M-Business projects.

The calculation of ROI is highly application dependent, process dependent, and user dependent. Internal initiatives are typically far easier to quantify than external initiatives. In general, it is easier to measure the cost taken out of the equation by increasing productivity for employees than to measure the influence one additional application or communications channel may have on overall customer behavior. The return on investment for customer-facing application can take on so many forms: increased customer acquisition; increased revenues per user; increased customer loyalty; increased customer retention; improved customer satisfaction; improved brand recognition; and positive word-of-mouth.

Despite these variables, it is often possible to estimate the return on investment prior to an enterprise M-Business initiative and to come up with useful numbers that are supportive of the project. Even with high-level calculations, the return on investment for many projects is compelling and jumps out from the numbers. Further levels of detail often serve to improve the overall value proposition as opportunity for revenue generation and resource cost reduction are added to the initial figures for productivity improvement.

Calculating the Return on Investment

To give an example of a return on investment calculation for an internal enterprise initiative, we'll take a real-life situation in the construction industry. In this example, the company wanted to increase the productivity of its field force in performing interior and exterior building restoration services to major clients. By automating many of the manual processes for the field force, the company planned to generate a significant return on investment by increasing productivity, reducing costs, and generating more billable hours. Additionally, the company aimed to increase customer satisfaction and retention by providing timely, accurate, and comprehensive information to their clients who were often off-site.

Table 6-1 shows some of the costs that went into the ROI projections and some of the savings that were expected.

Table 6-1 Example ROI Estimation for Field Force Automation in Construction Industry

Cost	Per Person	10	50	100	200	400
Devices	$2,000	$20,000	$100,000	$200,000	$400,000	$800,000
Wireless Service	$600	$6,000	$30,000	$60,000	$120,000	$240,000
Software & Services (Variable)– Model B	$2,000	$20,000	$100,000	$200,000	$400,000	$800,000
Software & Services (Fixed)– Model A	$250,000	$250,000	$250,000	$250,000	$250,000	$250,000
Total (Model B)	$4,600	$46,000	$230,000	$460,000	$920,000	$1,840,000
Total (Model A)	$252,600	$276,000	$380,000	$510,000	$770,000	$1,290,000
Cost Savings Productivity Increase	5%					
Hourly Rate	75					
Hours	2080					
Productivity Factor	50%					
Hourly Bill Rate	150					
Savings Amount	$7,800	$78,000	$390,000	$780,000	$1,560,000	$3,120,000
Revenue Amount	$7,800	$78,000	$390,000	$780,000	$1,560,000	$3,120,000
Total Amount	$15,600	$156,000	$780,000	$1,560,000	$3,120,000	$6,240,000
ROI (Model B)	339%	339%	339%	339%	339%	339%
Savings (Model B)	$11,000	$110,000	$550,000	$1,100,000	$2,200,000	$4,400,000
ROI (Model A)	6%	57%	205%	306%	405%	484%
Savings (Model A)	-$237,000	-$120,000	$400,000	$1,050,000	$2,350,000	$4,950,000
ROI (Model A) in Months	194.3	21.2	5.8	3.9	3.0	2.5

The ROI calculations in Table 6-1 take two different scenarios for estimating the cost of the software and integration services. One scenario, model A, assumes a fixed cost of $250,000 regardless of the number of end users the solution is rolled out to. The other scenario, model B, assumes a variable software and integration cost of $2000 per employee. In reality, these costs will depend upon the license fees of the software vendors (and whether they charge per server or per user, or combinations thereof) and also the amount of integration costs incurred. These integration costs can vary tremendously based upon the amount of work required. Integration into a large number of back-end systems will obviously be far higher than a package implementation that requires little or no integration with other applications.

Additional assumptions in the model included the following:

>> **$2000 Device Cost Per User**—For "ruggedized" handhelds.

>> **$600 Wireless Carrier Cost Per User/Per Year**—Average of $50 per month.

>> **5% Productivity Increase**—Project managers using the wireless application were estimated to save 2 hours per week.

>> **50% Productivity Factor**—Project managers were given a 50% productivity factor for their "saved time," since not all time would be recouped as billable hours.

>> **$75 Hourly Cost**—This was calculated by including not just the project manager but workers who would be impacted by the productivity increase

>> **$150 Hourly Bill Rate**—The bill rate to the customer.

Since the scenario labeled model B used a fixed $2000 cost per user for the software and services, the return on investment came out as a constant 339% regardless of the number of end users using the system. The other scenario, labeled model A, used an overall fixed cost of $250K for software and services. Economies of scale came into play in this scenario and the return on investment went from 6% to 484% between 1 and 400 users. The breakeven point for this scenario came between 10 and 50 users. At 10 users, the ROI was not realized in the first year, but at 50 users the ROI was realized in under six months.

Figure 6-1 shows a graphical comparison of the productivity increase (cost savings and revenue generation) versus the costs for both model A and model B. At larger numbers of employees, the return on investment becomes very compelling.

Additional Costs

This example shows a return on investment calculation conducted to a moderate level of detail. Many of the finer details were included in the assumptions. For example, the $250,000 cost for "software and services" could take into account costs for a variety of software and even some hardware purchases. If the full technical architecture of the solution has been planned or is already known via the packaged software vendor, it is easy to itemize all the costs associated with hardware and software including servers, devices, digital cameras, PCs (for administration), cables, firewalls, routers, Internet connectivity, and so forth. The services component is highly variable based upon whether the solution is package or custom and the level of integration required. Incorporating functionality for security and offline data collection capability may also add cost to the software and services component, unless it is included in the packaged solution.

Figure 6-1 Example ROI Estimation for Field Force Automation in Construction Industry.

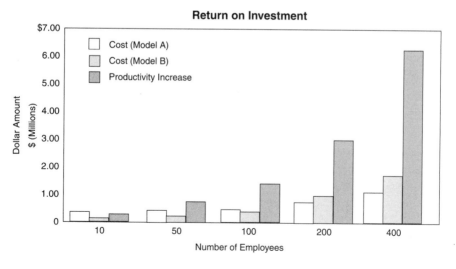

Finally, beyond the IT costs of implementing the solution, one has to look at the costs of on-going administration, maintenance, and support. This can include maintenance and upgrades to code, support for the IT infrastructure that runs the application, and application support for the end users themselves. We may reasonably assume that administration of the application itself increases productivity over the prior manual processes that the administrator had to perform.

Additional Cost Savings

Additional cost savings may be found in the form of cost savings owing to less paperwork and less costly communications. There may also be cost savings from reduced mailing costs if customer information is provided on an extranet site versus being printed and mailed to each customer. For some industries, this alone can yield several million dollars in cost savings per year or even per month.

Qualitative Benefits

It is often difficult to quantify customer-centric benefits, but it is worth adding all of these potential benefits to your overall business case. Some benefits may be so strategic to the enterprise, being well aligned with the corporate goals, that the project justifies itself on these merits alone. The initiative may also add the hard-to-quantify benefit of competitive advantage to the equation.

Net Present Value

Our return on investment calculations described above accounted for a single year scenario to provide some initial figures. We looked at the up-front one-time costs of hardware, software, and integration services, plus the recurring charges for wireless carrier service and the on-going percentage productivity improvements and revenue generation capabilities.

In many cases, it may be important to look beyond the one-year figures and look at the lifetime value of the project and the worth of the project in comparison to other enterprise initiatives that also require funding decisions to be made.

Another way to value M-Business initiatives is the Net Present Value (NPV) technique. Net Present Value takes into account the time value of money. That is, a dollar received today is worth more than a dollar received at any point in the future. NPV takes into account future cash flows and discounts them to their value today. The technique can be especially useful when comparing several M-Business initiatives and choosing which have the best investment value for the enterprise.

Note that NPV will typically only value hard-benefits (such as revenue generation or cost reduction) and not the soft benefits (such as customer satisfaction), which can also be a core part of the business strategy. One way to incorporate soft benefits into the NPV calculation is to assign relative dollar values to these soft benefits and add them to the cost savings and revenue generation numbers for an overall enterprise dollar benefit.

The NPV formula allows a risk factor to be assigned that will give a project a mathematical weighting based on its level of risk in terms of success or failure. This risk factor is actually the discount rate in the following formula for NPV:

$$NPV = (\$ \ in \ Future) \ x \ (1 + Discount \ Rate)^{-Number \ of \ Periods}$$

or

$$NPV = Future \ Value \ x \ Discount \ Factor$$

These risk factors are highly subjective, but the general rule is that a lower discount rate means a project with less risk. Thus, an M-Business project with a discount rate of 10% is far less risky in terms of success than one with a 30% discount rate.

If we now apply the NPV formula to our field force automation scenario, we can find out the net present value for the project. We'll assume a discount rate of 20% and a lifetime of the project of three years. We'll also take the numbers from the 100-person calculation and use just the cost savings numbers for productivity not for revenue generation opportunities. Additionally, we'll assume the $250,000 for software and services includes a $50,000 per year on-going fee for upgrades and support. Adding this $50,000 per year figure to the $60,000 per year figure for the wireless carrier service for 100 employees gives us yearly ongoing costs of $110,000.

Table 6-2   Example NPV Calculation for Field Force Automation in Construction
Industry

Date	Cost	Cost Savings	Future Value	Discount Factor	NPV
Start	$400,000		- $400,000	1.0	- $400,000
Year 1	$110,000	$780,000	$670,000	0.83333	$558,331
Year 2	$110,000	$780,000	$670,000	0.69444	$465,275
Year 3	$110,000	$780,000	$670,000	0.57870	$387,729
					$1,011,335

From Table 6-2, we arrive at a Net Present Value for the project
of $1,011,335. Any NPV value above zero provides a return on
investment, so this project is a very strong candidate for deployment.
The calculations can be further refined by adding in additional cost
factors for on-going end user support and training, and on-going IT
management of the application. The example above does not aim to
be completely exhaustive in terms of costs factored in, but aims to
demonstrate how NPV can be calculated and used as an additional
investment metric for M-Business projects.

Determining the Discount Rate

Since the discount rate is highly subjective, it is worth noting some of
the factors that can affect the discount rate (i.e., the risk factor) when
estimating net present value for M-Business initiatives. The following
items are examples of risk factors to consider within an individual
project:

>> Anticipated lifetime of the application

>> Risk comparison with other IT initiatives

>> Anticipated learning curve for end users

>> Change readiness of the organization

>> Company history with prior adoption of emerging technologies

>> Wireless coverage in area of business activity

>> Anticipated application support and service costs

>> Wireless package vendor maturity and product maturity

>> Ability to leverage hardware investment with other M-Business initiatives

>> Accuracy of productivity improvement projections

>> Accuracy of revenue generation and cost savings projections

>> Accuracy of process improvement projections

The discount rate should be estimated with awareness of current discount rates in use within the organization—particularly the IT department, if NPV calculations are a part of the decision-making process. The discount rate should be higher than typical IT application deployments owing to the relative emerging technology status of M-Business applications. The risk of implementation is higher than a standard E-Business application owing to the fact that the enterprise is dealing with new devices, new service providers, new application software and vendors, and new business processes. Additionally, all of the barriers to adoption that were discussed earlier in the book such as limited bandwidth, small form factors, poor coverage, and emerging standards are in effect to varying degrees of severity per application and per enterprise.

Business Agility Lessons

M-Business Strategy

>> An assessment of the current state business and technology environment can ensure existing competencies are best leveraged by M-Business initiatives.

>> An understanding of future business and technology goals and objectives can help to ensure alignment of M-Business initiatives.

>> Areas of opportunity may well lie across all user constituencies.

>> M-Business initiatives should be prioritized based upon return on investment and synergy with corporate objectives.

>> Initial applications should be self-justifying but should also serve to lay the groundwork for future applications to build upon.

Return On Investment

>> ROI can often be calculated by estimating productivity improvements alone.

>> Additional cost savings and revenues may be found in the form of less paperwork, reduced mailing costs, less costly communications, opportunities for additional billing, and reduced or eliminated administrative functions.

>> Qualitative benefits should also be articulated in the business case.

>> Net present value takes into account the time value of money. The discount rate is a factor of the risk of the project.

$$NPV = (\$ \ in \ Future) \times (1 + Discount \ Rate)^{-Number \ of \ Periods}$$

or

$$NPV = Future \ Value \times Discount \ Factor$$

7

M-Business Architectural Frameworks

As IT departments have moved from a cost center to a strategic partner to the business over the past twenty-five years, so have their applications. Just within the last five years, we have seen departmental client/server and Web applications grow to become business-critical enterprise applications. With so much of the business riding on these applications, corporations are constantly revisiting their E-Business architectures looking for optimal code reuse, business performance, and flexibility. In fact, some corporations view E-Business architecture as a strategic competitive advantage—their business, and the service offered to customers, is differentiated by their leverage of emerging technology, and the strength and flexibility of their underlying technical architecture.

The Case for Technical Agility

Traditionally, when we think of IT architectures we think of the requirements the architecture meets in terms of performance, scalability, reliability, maintenance, manageability, and security. An architecture is deemed successful if it meets the service level agreements for these attributes together with strong adoption and usage by end users. These architectural attributes come up again and again on every IT project, but with different priorities. In the financial services world, the most important attributes of an application are often performance and reliability. In other industries, there are often less stringent requirements for performance and reliability, and often other requirements become more critical to the overall success of the application. For example, in B2C E-Commerce, scalability can become a large factor—although somewhat intertwined with reliability, since reliability is impacted if the limits of scalability are exceeded.

In working with enterprise clients over the years, and especially in the E-Business era, I have seen two new architecture requirements that have emerged: those of flexibility and adaptability—in order words, technical agility. In the first few years of the E-Business era, many large enterprises built application upon application as quickly as they could with the intention of being first to market and competing on Internet time. While this brought their products and services to market quickly, and provided their customers and business partners with an electronic channel, it often had a negative impact on the overall IT architecture inside the enterprise. The applications that were built were often hastily assembled and were too rigidly defined to allow for future changes in direction within the applications. Thus they often could not respond to business events and opportunities.

In effect, the IT department, driven by the business conditions at the time, had made a tradeoff and had chosen speed over flexibility. Speed of development was favored over runtime flexibility. Short-term gain was placed in favor of long-term potential. This can be likened to the metaphor of constructing a building. If the building is constructed too quickly, and many additional structures are added as an afterthought, the building becomes compromised and the foundation is less solid. More up-front design for the accommodation of future growth can enable a sound foundation that can incorporate future additions to the structure at minimal cost.

The new breed of both E-Business and M-Business applications need to have technical agility engineered in. Flexibility and adaptability are perhaps the most important attributes in the new breed of IT architecture. It is important to add that enterprise IT standards, including adherence to "open" industry standards, are a critical mechanism for achieving this flexibility and adaptability.

Business Process Discovery

Technical agility can also be assisted by things as mundane as good documentation—both human-readable and machine-readable. To make a change to an existing application, developers need to know how and why the application was written the way it was in the first place. Documentation within the code is often the best form of assistance for developers working on previously-written code. One of the interesting areas within the enterprise application integration (EAI) arena right now is the ability of software to expose its application programming interfaces (APIs) via Web services. This leads toward the holy grail of EAI, where business processes between companies can be automatically discovered via the interfaces that they expose over the Web. Note that several have argued, rightfully so, that for two companies to integrate their systems it takes a lot more than technology—it also takes an existing business relationship and trust, so this holy grail may simplify the lives of developers, but it does not solve the entire issue of business-to-business integration.

An interesting twist to this EAI topic of business process discovery is that as we rely more and more on computers to help us with our integration tasks between disparate systems, we can also require them to document the interfaces that they discover as they go. For example, an enterprise application integration platform that integrates at the user-interface or screen level could document the interfaces it encounters as a non-technical end user drives the application through a set of screens in order to perform a transaction. After one iteration, the EAI tool would be able to produce a machine-readable API for how to perform the transaction. That is to say, it would be able to document the application it is integrating with so that a business scenario is rapidly turned into a callable transaction. For monolithic applications that can only be integrated with at the user-interface level, the ability to generate the API code automatically can be highly valuable.

The developments in enterprise application integration are important to M-Business architecture because M-Business applications often serve to extend the reach of corporate applications. EAI platforms thus help form the linkage between the wireless applications and the corporate systems they wish to access.

M-Business Technical Architecture

Views of M-Business Architecture

M-Business architecture, like E-Business architecture, is both an art and a science. Its definition and objectives depend upon the viewpoint of the individual practitioner. End users, business analysts, network architects, Web architects, data architects, and application architects will all have different views into the overall architecture and have different objectives. There are network architectures, data architectures, application architectures, systems architectures, component architectures, and so forth.

Very formal treatments of software architecture design and model-driven development can be applied using tools and methodologies from companies such as Rational Software. The company has a "4+1" model of architecture, which consists of a use-case view, logical view, process view, implementation view, and a deployment view. This "4+1" architecture view can be mapped to views within the Unified Modeling Language (UML), such as the use case diagram, class diagram, sequence diagram, package diagram, and deployment diagram. As described by Rational, "The Unified Modeling Language (UML) is considered the de facto standard object modeling language in the industry. The UML is the evolution of early approaches to object-oriented analysis and design."

In this chapter, I take the liberty of adopting a less formalized view of M-Business architecture with the goal of presenting some of the key considerations. I have done so in a manner that is digestible to business readers in addition to more technical audiences.

There are many ways we can represent M-Business and E-Business architectures diagrammatically—we can view the architecture from a logical perspective, a scenario-driven perspective, or a physical perspective to name just a few possibilities.

>> **Logical Views**—Logical views often present the high-level building blocks within the architecture or specific applications being described. They can be used to present the various application services that operate within the end-to-end enterprise architecture or specific application.

>> **Scenario Views**—Scenario views add an element of process and sequence to the logical views. They can be built using the logical components from the logical view with the addition of numbered arrows connecting the components in an ordered sequence as an application scenario is executed. A typical application example that we could describe with a scenario view might be a field service representative using a PDA device to access equipment maintenance information from back office systems over a wireless network.

>> **Physical Views**—Physical views can be used to translate the logical application services into specific hardware and software configurations that comprise the solution. It is here that the specific vendor products to be used in the solution are described and the exact specifications of the system can be documented.

The logical views and scenario views are excellent for discussing M-Business architecture with end users. They present a modular view into the services and components that comprise an application and the direction of flow between these components. In our discussions in this chapter, we will use logical and scenario diagrams in order to provide a view into M-Business architecture that is accessible to both IT and non-IT stakeholders.

Reference Architecture

The M-Business technical architecture for the enterprise should be viewed as a seamless extension of the current enterprise E-Business technical architecture. Ideally, it should provide support for wireless access to enterprise applications and data via any device, at any time, from any location, and by any user. User access should be determined by the role of each specific user and their associated access permissions for given applications.

A reference architecture for M-Business should include support for both the development and deployment of wireless applications. It should provide support for all wireless (and non-wireless) technologies and standards the enterprise deems to be important for future business benefit.

The reference architecture should enable wireless access by employees, customers, suppliers, and business partners in a holistic manner.

Figure 7-1 shows a logical M-Business architecture for an enterprise. As you can see, it is simply an extension of an E-Business architecture with additional application services in order to support wireless access. These additional services include gateway services such as WAP servers, communications services such as unified messaging, and integration services such as wireless middleware.

The figure shows five layers of functionality from the user interface layer to the data layer going from left to right. Within each layer, there are several modules that provide application services. Within the enterprise services layer, there are modules for knowledge management, customer relationship management, supply chain management, partner relationship management, enterprise resource planning, and so forth. The five layers are defined as follows:

User Interface

>> **Presentation Services**—This layer provides the user interface on the client device. The device may be a pager, PDA, cell phone, laptop, PC, fax machine, or any other communications device. Additionally, this layer includes the software functionality that presents the information on the device (including navigation), and potentially stores and manipulates data on the client. Of course, in an offline scenario, the device may actually contain components of user interface, business logic, and data all on the client device.

Business Logic

>> **Communication Services**—This layer provides the communication services such as support for HTTP via a Web server, support for the Wireless Application Protocol (WAP) via a WAP gateway, support for e-mail, messaging, speech, and multimedia functionality. Additionally, it provides the basic administration of these services in terms of indexing, query, and reporting capabilities.

Figure 7-1 M-Business Logical Architecture.

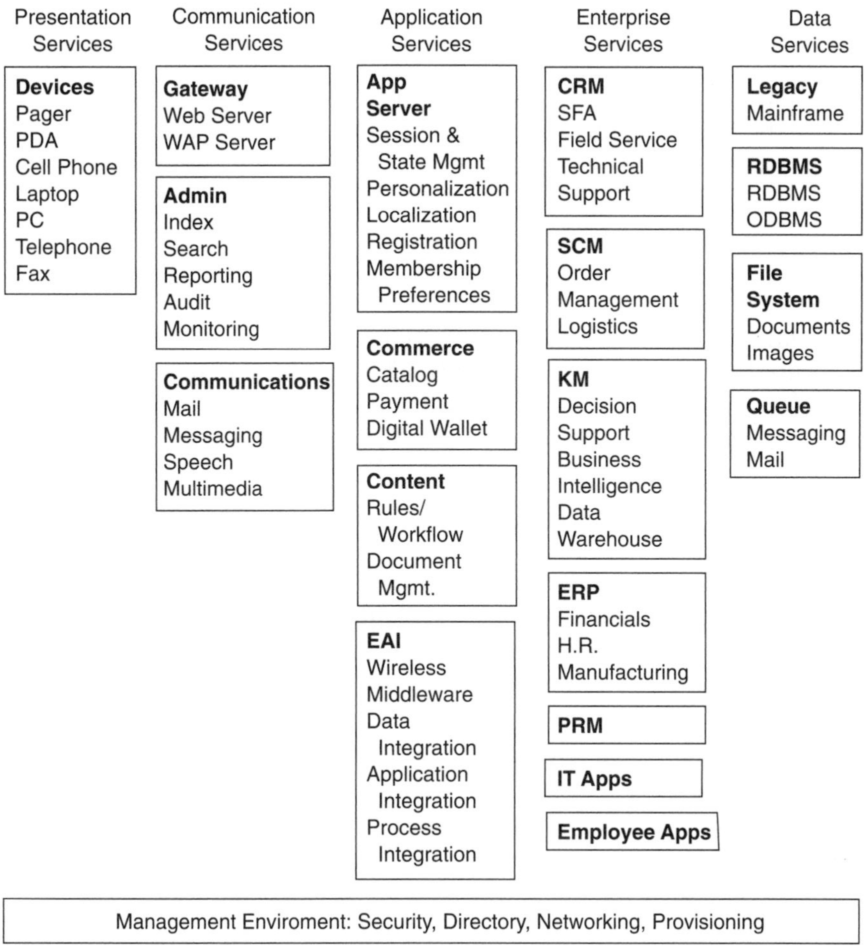

Presentation Services	Communication Services	Application Services	Enterprise Services	Data Services

Devices
Pager
PDA
Cell Phone
Laptop
PC
Telephone
Fax
	Gateway
	Web Server
	WAP Server
	
	Admin
	Index
	Search
	Reporting
	Audit
	Monitoring
	
	Communications
	Mail
	Messaging
	Speech
	Multimedia
		App Server
		Session &
		  State Mgmt
		Personalization
		Localization
		Registration
		Membership
		  Preferences
		
		Commerce
		Catalog
		Payment
		Digital Wallet
		
		Content
		Rules/
		  Workflow
		Document
		  Mgmt.
		
		EAI
		Wireless
		Middleware
		Data
		  Integration
		Application
		  Integration
		Process
		  Integration
			CRM
			SFA
			Field Service
			Technical
			Support
			
			SCM
			Order
			Management
			Logistics
			
			KM
			Decision
			Support
			Business
			Intelligence
			Data
			Warehouse
			
			ERP
			Financials
			H.R.
			Manufacturing
			
			PRM
			
			IT Apps
			
			Employee Apps
				Legacy
				Mainframe
				
				RDBMS
				RDBMS
				ODBMS
				
				File System
				Documents
				Images
				
				Queue
				Messaging
				Mail

Management Enviroment: Security, Directory, Networking, Provisioning

>> **Application Services**—This layer provides the core functionality to run E-Business and M-Business applications. Included in this layer are application server technologies, enterprise application integration tools, content management systems, E-Commerce and M-Commerce systems, and other software infrastructure functions.

>> **Enterprise Services**—This layer provides the enterprise application functionality. This is the application functionality that is used and seen by end users of the E-Business applications. Typical applications include enterprise resource planning, customer relationship management, supply chain management, partner relationship management, knowledge management, and so forth. These are typically packaged applications that may or may not have wireless extensions.

Data

>> **Data Services**—This layer provides the structured and unstructured storage of enterprise data. This includes mainframe systems, relational database management systems, object databases, file systems, message queues, and any other storage mechanisms.

Distributed Operating Environment

>> Finally, the Distributed Operating Environment cuts across all five layers of the logical architecture and provides management services for the applications in terms of security, directory services, networking services, and application provisioning.

One of the reasons that M-Business needs to be viewed holistically within the overall enterprise E-Business architecture is that it cuts across all of these five layers within the logical model. M-Business applications consist of components at the user interface level, business logic level, and data level. In a typical M-Business application, you may leverage specialized components that are built solely for wireless, plus you may leverage components that are M-Business-neutral and which apply equally well to both wireline (wired) and wireless applications.

For example, a wireless extension to an existing Web application may take the existing HTML content, transform it to XML, and allow the developer to convert the XML content to various output formats with the aid of style sheets for use with PDAs and cell phones. In this case, the presentation logic has been changed but the business logic and data layers remain unchanged; as such, they can be leveraged from the initial application with minimal changes.

Architectural Principles

The following are some architectural principles that may help in setting the technical direction for M-Business applications:

Flexibility

>> Support published and accepted standards such as XML, XSL, HTML, HTTP, WAP, WML, and SMS

>> Support various networks and protocols

>> Support both voice and data

>> Support existing enterprise operating systems

>> Support offline and online usage via synchronization

>> Support as many mobile devices as necessary and include mechanisms for support of new devices as they appear

>> Support integration with application servers, enterprise application integration technologies, and packaged and legacy applications

Availability/Reliability

>> Applications should support high-availability via load balancing, fault-tolerance, and fail-over

Manageability/Maintainability

>> Applications should be component-based in order to support reuse across applications

>> Application development and deployment environments should be manageable via an integrated graphical user interface or hooks to other application development or systems management environments

>> Applications should be non-invasive and minimize impact on existing applications, infrastructure, and networks

Performance

>> Applications should support enterprise requirements in terms of end-to-end response times

Scalability

>> Applications should be scalable to meet the needs of the enterprise in terms of number of end users and the number of simultaneous transactions

Security

>> Applications should support various levels of security and include support for authentication, access control, and encryption

>> Applications should integrate with existing security frameworks such as PKI

Usability

>> Applications should be developed to meet the usability and user interface requirements of end users

As you can see, most of these architectural principles apply equally well to E-Business applications as to M-Business applications. The main point is to ensure that applications support the wide range of standards, protocols, networks, and devices that will be encountered in real-life deployments. This is especially true when deploying applications for use outside the enterprise, where these factors cannot be as easily standardized or controlled.

Enterprise Architecture

The enterprise M-Business architecture components consist in general of a wireless middleware platform implemented alongside other enterprise servers, such as Web servers and application servers.

Wireless middleware platforms, or wireless application gateways (WAGs) as they are also known, are the most common approach to enabling E-Business applications for wireless access. Rather than rewriting applications to support mobile devices (as in custom approaches), these wireless middleware platforms let developers

import existing ASP, HTML, JSP, and other files into a common system with a single method of adding XML tags. Those tags describe
how elements of the Web page should be presented on different wireless devices. For those of you new to some of these acronyms, a quick
explanation is in order. ASP and JSP stand for Active Server Pages and
Java Server Pages, respectively. They are simply files that reside on the
server-side of an application and contain both presentation code such
as HTML together with business logic, which is processed by the
server before returning the output results combined with the HTML
back to the client device.

There are a large number of software companies who provide wireless middleware platforms and solutions for the enterprise. A brief listing of some of these companies is as follows: @Hand, 2Roam, 724
Solutions (which merged with Tantau), Aether Systems, Air2Web,
AlterEgo Networks, AvantGo, Brience, Capslock, Everypath, Extended
Systems, IBM, iConverse, InfoWave, JP Mobile, Microsoft, MobileQ,
Oracle, Seagull Software, ViaFone, WirelessKnowledge, and Xiam. As
you can see, the market is comprised of a combination of major software companies such as IBM, Microsoft, and Oracle, together with
many new entrants who specialize in the wireless market.

Most of these software vendor products support open or industry
standards including XML, XSL, HTML, HTTP, WAP, WML, and
SMS. Additionally, some vendors have developed their own proprietary extensions in order to increase the functionality of their products or the ability to integrate with their products.

From an architectural perspective, some of the most important
features to look for and questions to ask when selecting a wireless
platform are as follows:

>> **Application Services**—Does the platform provide horizontal applica-
tion functionality such as e-mail and personal information management functionality, or vertical application functionality specific to
your business requirements? If the platform does not provide these
features, does it provide the necessary hooks into other such applications such as Microsoft Exchange and Outlook or Lotus Notes?

>> **Content Management**—Does the platform provide functionality
for content management? This includes the ability to adapt content based upon specific output devices and network bandwidths.

>> **Connectivity Management**—Does the platform provide connectivity to the required communications networks and protocols? This can include the ability to connect with carrier networks, wireless gateways, Web servers, and proxy servers. Many wireless middleware platforms incorporate their own wireless gateway for conversion between WAP and HTTP protocols.

>> **Integration Services**—Does the platform provide the ability to integrate with external systems such as enterprise ERP, CRM, SFA, and SCM packages plus other back-end systems including legacy mainframes and database servers?

>> **Performance Management**—Does the platform provide performance management capabilities that include load balancing, fault-tolerance, fail-over, and content optimization based upon device and network capabilities?

>> **Security Services**—Does the platform provide the relevant security services for user authentication, access control, and encryption?

>> **Application Management Services**—Does the platform provide the necessary application management services such as personalization, session and state management, preference storage, and profile management? Additionally, it is important to determine how the platform supports the on-going management of applications and users once the solution is deployed. How can application changes be pushed down to devices that operate in an offline scenario? How can user access to applications be issued, renewed, or revoked? How can applications be removed from devices?

>> **System Management Services**—Does the application provide system management services such as server administration, usage logging, provisioning, reporting, and billing? This is especially important for carrier-based or communications service provider deployments, where users need to be provisioned and billed for their usage of M-Business applications.

>> **Development Environment**—How easy is the development environment to work with? What is the learning curve? Does it work with open or proprietary standards? Does it incorporate a graphical development environment or does a lot of code need to be manually written in order to develop wireless applications?

How does it integrate with other development environments from major software companies? How intelligently does it pull existing Web content into the environment for adaptation to wireless devices?

These nine features can be considered the core features inherent in wireless middleware. Whichever vendor product you look at, you'll most likely find some or all of these services within the platform. They will often have slightly different names attached to each service, but will have the same or very similar functionality underneath.

Enterprise Mobile Server Architecture Example

JP Mobile
http://www.jpmobile.com

As an example of an enterprise mobile server architecture, we'll take a look at the software platform offered by JP Mobile—one of the players in the enterprise mobile server and wireless application service provider marketplace.

JP Mobile is a privately held company based in Dallas, Texas. The company was founded in 1995 and currently develops and markets the JP Mobile SureWave Mobile Server. The product is provided primarily as licensed software, but is also available as a hosted service. The high-level architecture of the platform is shown in Figure 7-2.

In the figure, you can see that the JP Mobile SureWave Mobile Server can be implemented behind an enterprise firewall or can be provided as a hosted service. The three major categories of service provided by the platform include mobile services, core services, and enterprise services. Within the enterprise services area, the platform connects to the enterprise data center and target enterprise applications such as Siebel and I2 via interface adapters that support standards such as XML, HTTP, JDBC, JMS, RMI, CORBA, cHTML, and VoiceXML. Within the mobile services area, the platform connects to the mobile devices such as pagers, PDAs, and WAP phones via the network gateway that supports standards such as SMS, WAP, and Bluetooth.

Any application an enterprise desires to have mobile access to can sit on top of the mobile server. JP Mobile has two packaged applications, SureWave Mail, which gives enterprise users mobile access to Lotus Notes, Exchange and IMAP4 mail servers and SureWave Web, which allows Internet and intranet information to be extended to wireless devices.

Figure 7-2   Enterprise Mobile Server Architecture. Source: JP Mobile, Adapted.

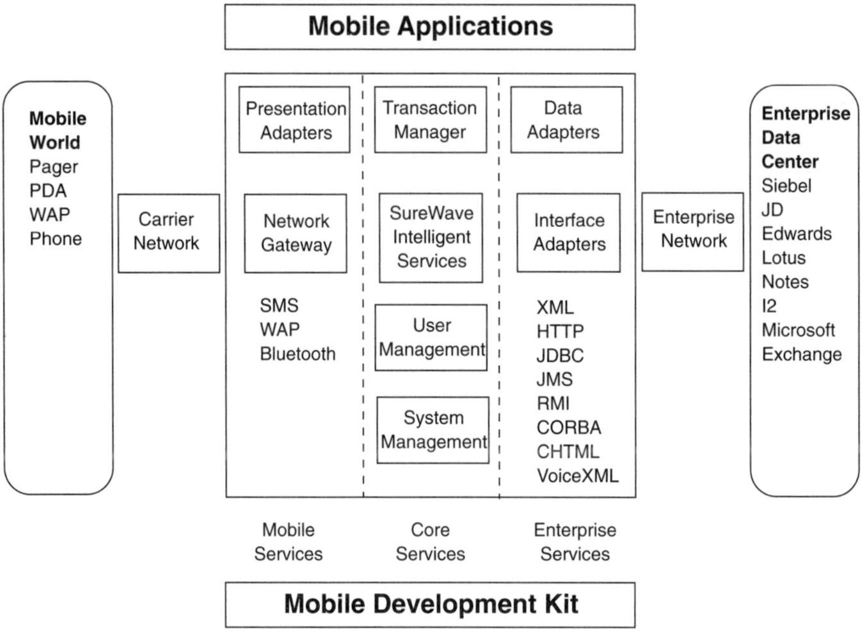

Sales Force Automation Example

The above diagram shows a typical architecture for an M-Business sales force automation scenario as a subset of our earlier enterprise M-Business logical architecture diagram.

This example assumes that the application resides on the server-side within the enterprise and is accessed over a wireless connection from the user's device in real-time. This wireless case is a slightly different scenario from what we might call the mobile case. In the mobile case, users often have applications fully loaded on their handheld devices for later synchronization with the enterprise when in a suitable wireless coverage area. This distinction needs to be made because the application architectures are different in these two scenarios.

In the scenario illustrated on the next page (Figure 7–3), a sales person using a personal digital assistant (PDA) device equipped with a wireless modem or WAP-capable cell phone is able to access a sales

Figure 7-3 Sales Force Automation Example.

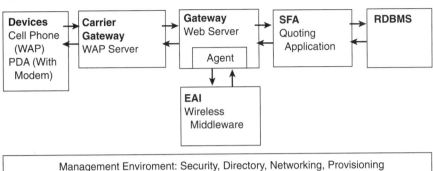

force automation application residing within the enterprise. A typical transaction may be that of providing a quote for a customer. The PDA device or WAP phone communicates with a carrier's WAP server acting as a gateway. Here's one example of how the application might work:

1. The sales person completes the fields within the quote form on the PDA device or WAP phone and submits the form by tapping or pushing a submit button.

2. The request for the quote is sent from the client device to the WAP gateway within the carrier network via the WAP protocol over the carrier's wireless network.

3. The carrier WAP gateway then directs the request to the Web server within the enterprise via the HTTP protocol over the Internet.

4. A software agent (part of the wireless middleware) running on the Web server reads the incoming request, recognizes the request is coming from a wireless device, and redirects the request to the wireless middleware application.

5. The wireless middleware application substitutes the requested filename with a different filename that has been formatted for the specific user device.

6. The wireless middleware passes the new filename back to the Web server for processing.

7. The Web server processes the filename by redirecting the request to the sales force automation (SFA) application.

8. The SFA application takes the input fields from the form, calculates the quote, and passes the results back to the Web server. It accesses a relational database in order to determine appropriate pricing for the specific customer receiving the quote.

9. The Web server sends the response to the WAP gateway using the HTTP protocol over the Internet.

10. The WAP gateway sends the response back to the client device using the WAP protocol over the carrier's wireless network.

11. The client device receives the response and displays the final quote on the user's PDA device or WAP phone.

As you can tell, the sequence of processing can get quite complex with numerous servers involved in the transaction. However, this is only slightly different from typical E-Business application scenarios. The main difference is that a WAP gateway is involved in order to translate the HTTP protocol to the WAP protocol for delivery over the carrier's wireless network. Additionally, there may be a software agent (which is part of the wireless middleware platform) running on the Web server to intercept incoming requests from wireless devices and re-route the processing to allow for the substitution of device-specific content.

Carrier and Wireless Application Service Provider Architecture

When considering M-Business applications targeted at the enterprise, the carrier and wireless application service provider architectures are essentially similar to the in-house enterprise architectures. The major difference is that they are often built with increased capabilities for scalability and performance and have added features for multi-hosting, provisioning, and billing. Wireless application service providers essentially perform an outsourced service for wireless enablement of enterprise applications. The carriers provide a variety of wireless data services from simple e-mail, personal information management, and unified communications services to wireless application services that are similar to those offered by the wireless ASPs. In this section, we'll look briefly at some of the architectural components that comprise both of these wireless data services. For enterprises that pursue an outsourced approach to some of their M-Business applications, an understanding of the high-level details surrounding these architectures can help in determining the range of potential opportunities.

Wireless Application Service Provider Architecture

The wireless application service provider's function is either to provide simple self-contained applications such as personal information management or to provide the wireless middleware services necessary to wirelessly enable back-end enterprise applications such as customer relationship management, sales force automation, field force automation, supply chain management, and enterprise resource planning. In order to access these back-end applications that reside within the enterprise, the wireless application service provider's often provide an API in the form of an XML-based interface for enterprise developers to hook into. An example is the XML-based API provided by Air2Web, a privately-held wireless application service provider based in Atlanta, Georgia.

When selecting a wireless application service provider and their technology platform, many of the same wireless middleware questions that we looked at earlier within the Enterprise Architecture section still apply. The only major difference is to determine your companies philosophy toward insourcing versus outsourcing. The question to ask is whether your M-Business applications will be considered strategic to corporate objectives or not. If they are a strategic weapon for the enterprise to achieve business goals, then you may wish to keep the technology inhouse. The other question to ask is whether or not this is a core competency you wish to develop and maintain inhouse. Some companies such as ADC Telecommunications, which we covered in our earlier case study, consider their M-Business applications strategic but decided to outsource the solution since it was not a core competency they wished to maintain. They also gained a time-to-market advantage versus implementing the solution themselves. Of course, some vendors such as JP Mobile will provide the option of both licensed software or a hosted service, so you will be able to pick the situation that is most appropriate to your needs.

Carrier Architecture

The wireless carriers often provide data services such as unified messaging, a wireless data portal, and wireless applications to their subscribers. We can expect to see more offerings from the carriers as time goes by and as they move from voice to value-added data services as discussed in Chapter 2.

To illustrate the architectural components that a wireless carrier may include in their current and future wireless data services, we'll take a look at the platform from Openwave Systems.

Carrier-Based Architecture Example

Openwave Systems Inc.
http://www.openwave.com

Openwave Systems, Inc. is a publicly traded software company (NASDAQ: OPWV) formed by the merger of Phone.com and Software.com. The company provides software for communications service providers, including wireless and wireline carriers, in order to build boundary-free, multi-network communications services for their subscribers. The Openwave Mobile Browser is embedded in more than 70% of all Internet-enabled phones.

Openwave's Services OS framework is the companies' open, Internet Protocol-based software platform for next-generation data services. It includes three core layers of functionality: mobile services infrastructure, platform services, and communication services. Some of the carrier-specific functionality within this framework includes the unified messaging, portal framework, provisioning manager, and mobile access gateway. The Services OS framework is shown in Figure 7-4.

As shown in the figure, the communication services layer provides many functions for e-mail and messaging. The unified messaging solution provides access to e-mail, voice mail, and fax over any device. For example, a user can dial in to their account, listen to e-mail over the phone, and reply to e-mail by voice. They can also forward voice mail to any e-mail address or listen to voice mail in the context of the Web session. The VoiceXML architecture allows for customization of voice menus and voice-driven applications.

Within the platform services layer, some of the components such as the portal framework, download fun, and provisioning manager are fairly specific to the wireless carriers in targeting consumers, but the components that are of perhaps greater interest to the enterprise are the location service broker and the enhanced services framework. The location service broker accepts location information from either handset-based or network-based systems and then makes it available via a well-defined interface for developers. It supports both subscriber-initiated requests or network-initiated requests.

Figure 7-4   Openwave Services OS Architecture.

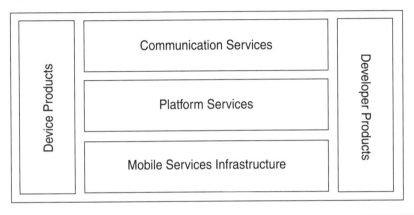

It is important for an enterprise to understand the high-level aspects of the communications service provider architecture because in certain instances you may need to integrate with them and subscribe to their services, which may be otherwise unavailable. Services such as location determination may not be available other than via this type of platform, since the wireless carrier owns the location information. By integrating directly into this platform or into software from an independent software vendor who has already made the integration and has added a layer of application functionality, you can take advantage of location-based services for your user constituencies and assets.

A lot of questions remain to be resolved around the business model and technical logistics of location-based services. One of the questions is how the carriers will exchange location information between one another as corporate users and assets roam from one carrier network to another. Beyond these questions relating to location-based services, there are also many questions outstanding as to the role the carriers will ultimately play in the entire M-Business value chain. With a larger and larger percentage of subscriber revenues predicted to gravitate toward value-added data services, it will be interesting to observe which carrier-platform data services become most differentiated and valuable both for consumers and for the enterprise. As IT gravitates toward an always-on, network-centric model with Web-based services exposed for others to leverage, the carriers have an excellent opportunity to be a critical part of the value chain and to build their own ecosystems of partnerships and developer communities. By opening their networks so that developers can write to their application programming interfaces, the carriers get the opportunity to build best-of-breed applications to the top of their core platform architectures. This in turn will provide an additional layer of value for their subscribers.

Business Agility Lessons

M-Business Technical Architecture

>> IT architecture can yield strategic competitive advantage.

>> Technical Agility (flexibility and adaptability) are the new requirements for IT architecture in addition to the traditional requirements of performance, scalability, reliability, security, maintainability, and manageability.

>> M-Business architectures can be represented in logical views, scenario views, and physical views.

Enterprise Architecture

>> Wireless middleware platforms are a key component of M-Business architectures. They represent a core server technology that should be positioned alongside your Web servers and application servers.

>> They typically provide application services, content services, connectivity services, integration services, performance management, security services, application management, systems management, and a development environment.

>> The main objective of wireless middleware is to enable applications to be delivered over any device, any network and any content format.

>> Vendors include @Hand, 2Roam, 724 Solutions, Aether Systems, Air2Web, AlterEgo Networks, AvantGo, Brience, Capslock, Everypath, Extended Systems, IBM, iConverse, InfoWave, JP Mobile, Microsoft, MobileQ, Oracle, Seagull Software, ViaFone, WirelessKnowledge, and Xiam.

Wireless Application Service Provider Architecture

>> Wireless application service providers essentially perform an outsourced service for wireless enablement of enterprise applications.

>> When evaluating wireless application service providers, it is important to determine your companies' philosophy towards insourcing versus outsorcing.

Carrier Architecture

>> Carrier architectures include components for communications services, platform services, and mobile services.

>> Communications services include applications such as e-mail, instant messaging, and unified messaging.

>> Platform services include applications such as location-based services, data portals, synchronization services, and provisioning applications.

>> Carrier architectures are of importance to the enterprise because services such as location determination may only be available via this type of platform. The wireless carrier often owns the location information.

Diagnosing Your Business Agility

>> Does your current enterprise IT architecture provide the levels of business agility that you require both now and in the future?

>> Have you considered wireless middleware as a core technology within your enterprise IT architecture?

>> Have you considered the potential role of wireless application service providers and wireless carriers in your IT strategy?

8

M-Business Technologies and Implementation

This chapter deals with M-Business technologies and their implementation—where the rubber meets the road. Once an M-Business strategy has been formed and initiatives have been identified, justified, prioritized, and planned, technology choices can be made and custom or packaged applications can be implemented.

In this chapter, we look at some of the M-Business technologies available for enterprise deployments and some of the trends within the software industry that are shaping the vendor offerings in this space. One of these trends is the convergence of wireless middleware, application servers, and enterprise application integration. Another factor is the fact that vendors are converging on the wireless application space from a number of different entry points. These entry points include vendors coming in from the packaged application arena, the infrastructure arena, and the specialty wireless application arena. One of the outstanding questions for the enterprise is whether to pursue a vertical or horizontal approach to wireless enablement. The vertical approach generally means that the enterprise wirelessly enables applications by using the wireless extensions of the packaged application vendors. The horizontal approach generally means that the enterprise wirelessly enables applications by viewing wireless middleware as a key component of the IT infrastructure that needs to be plugged in alongside Web servers and application servers.

In addition to looking at some of these trends within the software industry, we'll also look at wireless middleware platforms and M-Commerce platforms, review some of the key features that these platforms provide, and provide some business and technical considerations when embarking upon application initiatives. In the M-Business Implementation section, we'll also look at some guidelines for performing vendor selections using a requirements matrix approach and some recommendations for implementing M-Business within the enterprise.

M-Business Technologies

As we saw in the chapter on M-Business Architecture, one of the main technologies for wireless implementations is the wireless application middleware platform. However, these platforms are evolving in the broader context of other enterprise software applications and infrastructure that is also gaining wireless functionality.

Convergence of Wireless Middleware, Application Servers, and Enterprise Application Integration

Wireless middleware platforms are one technology that the enterprise should consider as they start to implement their wireless strategies. Before deciding on a wireless middleware platform, keep in mind that we are likely to see application servers, enterprise application integration (EAI) technologies, and wireless middleware converge toward a common platform within the next 12 to 24 months. Figure 8–1 shows the convergence of these three product categories

All three of these software categories have mostly discrete functions today. The application servers function as the execution environment for middle-tier business logic, the integration technologies

Figure 8-1 Convergence of Wireless Middleware, Application Servers, and Enterprise Application Integration.

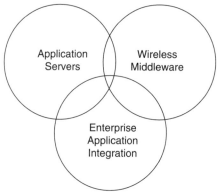

bring together a variety of enterprise and E-Business applications, and the wireless middleware platforms extend content from Web-based systems to wireless devices.

The three technologies are rapidly coming together as vendors merge, create partnerships, and introduce new products and solution sets. Examples include the participation of companies such as 2Roam, Aether Systems, AGEA, Brience, iConverse, and Volantis Systems in the BEA Wireless Star Solution program, which is part of BEA's Star Partner Program.

These new products or product combinations via partnerships effectively lead to a new breed of application server platforms that include built-in support for EAI and wireless middleware. Today, these features can be assembled with a variety of products—sometimes from the same vendor such as IBM, Microsoft, or Oracle. Soon, they may well be standard features in a single integrated platform. Until this happens, the IT department will be challenged with making disparate technologies work together to provide seamless wireless application solutions for their end users.

Of course, wireless applications for the enterprise are not only a technology-based decision. It is a decision that requires the unified collaboration of business and IT management in order to create solutions that meet business objectives and IT realities. What is clear is that the selection of wireless middleware needs to be made within this broader context of enterprise IT infrastructure, which includes both application servers and enterprise application integration technologies.

Software Vendors Are Approaching
M-Business from Many Different Angles

In addition to the macro-level convergence of the software categories of wireless middleware, application servers, and enterprise application integration technologies, we are also seeing a variety of software vendors coming into the space via their traditional backgrounds.

Some are extending their Web applications to the wireless domain such as the packaged application vendors for customer relationship management, supply chain management, enterprise resource planning, and sales force automation. Examples of these package vendors include i2, Oracle, PeopleSoft, SAP, and Siebel.

Others are approaching wireless from an infrastructure perspective: this includes vendors in the collaboration and groupware space, the content management space, the enterprise application integration space, and also the relational database management system space. Examples of these vendors include IBM/Lotus, Microsoft, Oracle, Vignette, Vitria, and WebMethods.

An entirely new category of vendors are providing specialty wireless applications, alerts and notifications, and wireless middleware specifically aimed at wireless deployments. Examples here include companies such as Antenna Software, CellExchange, EnvoyWorld-Wide, FieldCentrix, and all of the other wireless middleware vendors previously outlined.

Thus, the enterprise considering wireless deployment is hit from almost all angles with these varied options for wireless enablement. Figure 8–2 shows this spectrum of software vendors approaching the wireless arena.

One of the main questions is whether the enterprise should take a horizontal approach or a vertical approach to solving this problem. In the horizontal approach, the enterprise views wireless as an item within their IT infrastructure that needs to be plugged in to support any and all future applications. In the vertical approach, the enterprise

Figure 8–2   Software Alternatives for Wireless Enablement From a Horizontal and a Vertical Application Perspective.

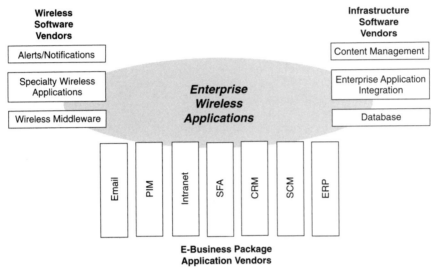

E-Business Package
Application Vendors

views wireless on an application-by-application basis using the wireless functionality provided by the package vendor.

In reality, the enterprise will most likely adopt both approaches. If the package application vendor provides robust wireless functionality, then this may often be sufficient and a more expedient and lower cost way to extend these applications to the field. If new wireless application development is required, or the enterprise requires greater wireless functionality than their package vendors currently support, then the wireless middleware approach is often appropriate.

Wireless Middleware Platforms

If wireless middleware becomes a part of your enterprise M-Business strategy, perhaps in addition to leveraging wireless extensions of packaged products such as Siebel and SAP, how should you go about revising your non-packaged applications to support both the wired and wireless worlds? Firstly, aim to build an adaptive architecture—one that provides enough flexibility to support today's devices and the new access devices, browsers, networks, and standards of tomorrow.

The available approaches for implementing your wireless middleware solution break down into three categories: fully-automated transcoding, configurable converters, and custom development. We could also term these three categories fully-automated, semi-automated, and manual approaches to wireless enablement.

>> **Fully-Automated Transcoding**—The transcoding technique simply takes existing Web content such as HTML and automatically converts it into other formats such as WML for cell phones. There are several problems with this approach. The interfaces are so different that the user experience will almost certainly be compromised.

>> **Custom**—The custom technique is the best approach for the user experience, but is also the most costly. Not only do you have to write code for the current multitude of WAP phones, PDAs, and other devices currently on the market, but custom code has to be written and maintained for new devices and standards that appear in the future. This can lead to a versioning problem in maintaining the code and in creating new code for your applications. It can be a useful technique, however, if you need highly specialized applications and you can limit the number of device types that you have to support.

>> **Configurable Converters**—In general, configurable converters present the optimal solution, since they provide the best method of supporting different devices and keeping an acceptable level of user experience—a development environment built around these solutions is absolutely essential. Most of today's wireless middleware platforms, as we learned in the previous chapter on M-Business architecture, fall into this category. They provide a development environment that will allow you to customize your content. Moreover, they also provide many services that can help to speed the development effort and provide a more robust and flexible deployment environment.

In the same manner in which a holistic enterprise application integration strategy can be adopted with solutions for integration at the user interface, method, application interface and data layers, a holistic approach to wireless translation that spans fully-automated, semi-automated, and custom techniques can also be adopted. In other words, you may find that your wireless middleware platform meets 80% of your wireless application needs, but that you still have to resort to custom programming in certain application instances.

The challenge for IT is to determine which applications and areas of functionality to wirelessly enable, which vendors to deal with, and how to deliver the wireless service—as a hosted service or as enterprise software.

Considerations When Choosing Wireless Middleware Platforms

To help you in selecting your wireless middleware platform, here are some business and technical considerations. Additionally, the next section of this chapter on the requirements matrix and vendor selection will provide a quantitative mechanism to help you in your decision making.

Business Considerations

>> Is there a clear business case and strategy around M-Business that will help to drive these technology purchases?

>> Do the immediate business needs justify the purchase of a wireless platform or is custom development sufficient?

>> Are there any cultural barriers to overcome in deploying the solution to end users?

>> Is sufficient change management in place for end users?

>> Are the scope and objectives of the initiative well understood?

>> Are the critical success factors well understood?

>> What is the track record of the software company and its products?

>> Does the vendor have a strong base of enterprise customers already using the product?

>> What are the terms and conditions surrounding the license agreement?

>> What is the product pricing?

Technical Considerations

>> Is the software customizable?

>> How scalable and secure is the platform?

>> What mechanisms does the vendor have for supporting new devices that appear on the market?

>> How conducive is the software to complex applications and highly transaction-oriented content?

>> How does the software deal with binary objects in addition to text-based files such as ASP, HTML, and JSP?

>> Does it support synchronization using standards such as SyncML?

>> Does it support voice access using standards such as VoiceXML?

>> Does the vendor have proprietary languages and APIs?

>> Does it interface with existing custom and packaged applications?

>> How well does it interoperate with application servers and with EAI software?

>> How well does it integrate with legacy and database systems?

>> How well does it support online and offline applications?

>> How well does it support application management and systems management functions?

>> Has the technology been evaluated against current IT standards and guidelines?

>> Does it meet IT architectural standards and service levels?

M-Commerce Applications

While wireless middleware provides support for multiple-carrier networks, multiple-wireless devices and multiple-content formats, M-Commerce software provides the added support for wireless transactions. These M-Commerce solutions often allow users to provide payment for purchases via a variety of mechanisms including prepaid accounts, credit cards, and telephone bills. As we saw in Chapter 2, trials and in some cases production implementations of M-Commerce services are underway that provide almost every permutation of payment mechanism and business model.

Software vendors in the space include 724 Solutions, Brokat Technologies, More Magic, Openwave, Qpass, Trintech, and several others. Some vendors are more carrier-centric in their approach such as Openwave, while others such as 724 Solutions are more focused on financial service institutions. Some of the features that you'll find within M-Commerce software include standard electronic commerce functionality that has been adapted for use over wireless devices. The functionality in these transaction platforms may include the following:

>> **Registration**—Ability for customers to register for M-Commerce services and to receive appropriate software or hardware in order to enable the transactions to occur.

>> **Personalization**—Ability to personalize content and pricing based upon individual user preferences and profiles.

>> **Product Marketing and Promotion**—Tools for marketing and promoting products, services, and content to customers.

>> **Catalog and Pricing**—Ability for customers to browse through a catalog of product, service, or content offerings and make their selections. Ability for operators, enterprises, or content providers to enter pricing information related to the content and to dynamically change pricing information.

>> **Search**—Ability for customers to search for specific product, service, or content offerings within the catalog.

>> **Payment Methods**—Ability for customers to pay for product, service, or content via a variety of payment mechanisms including prepaid accounts, credit cards, and telephone bills.

>> **Customer Account Management**—Tools for customers to manage their personal profiles and review their purchase history.

>> **M-Wallet**—An electronic account (a server-side digital "purse" or "wallet") suitable for small payments. These wallets typically operate on a pre-paid or post-paid basis and can be recharged via other payment methods. They store, manage, and transfer personal and payment information between consumers and merchants.

>> **Gateways**—Gateways to financial institutions for authentication and billing purposes.

>> **Security**—Ability to authenticate end users and provide encrypted communications with support for a variety of security standards and techniques such as username/PIN, digital signatures, digital certificates, WAP/WTLS, and HTTPS/SSL.

>> **Integration Tools**—Ability to integrate with existing legacy systems such as billing and banking systems and merchant catalogs. This may include tools for both M-Commerce platform operators and financial service providers and merchants wishing to integrate with the platform.

>> **Administration Tools**—Services for operation and maintenance of the M-Commerce transaction platform. Includes management tools for merchants or operators to provision applications, manage catalogs, and provide customer care.

>> **Development Tools**—Services for developers to build and customize the M-Commerce transaction platform.

The roles that are present within an M-Commerce transaction can include several distinct parties. The transaction may involve a customer, an M-Commerce platform operator such as a wireless carrier, and a merchant. Other roles may include authentication service providers and payment service providers. These roles may be provided by separate companies, each fulfilling a particular role, or they may occur in various combinations based upon the specific business model in question.

M-Commerce Transaction Example

Figure 8–3 shows a typical process flow for how an M-Commerce transaction may occur. In this example, the roles present include the customer, the wireless operator, the merchant, and the payment service provider. In this example, the wireless operator runs the M-Commerce platform and the payment service provider could be a credit card company, billing and clearing company, or other type of financial institution responsible for the financial aspects of the transaction. The payment service provider has issued a virtual wallet to the customer and stores their profile information including payment preferences and shipping address on their server.

The process flow for this M-Commerce transaction is as follows:

1. The customer selects items to purchase over a wireless device and proceeds to the merchant checkout.

Figure 8-3   M-Commerce Transaction Process Flow.

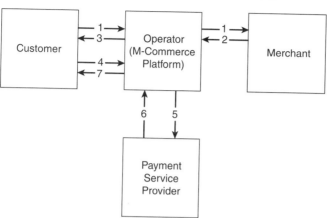

2. The merchant requests authentication.

3. The M-Commerce platform residing at the operator site recognizes the customer has a digital wallet and passes the authentication request to the customer.

4. The customer enters a PIN.

5. The M-Commerce platform passes the PIN to the payment service provider for authentication with the digital wallet.

6. The payment service provider authenticates the customer and passes the digital wallet information containing the payment method and shipping address back to the M-Commerce platform.

7. The customer is presented with the checkout screen and a "pay now" option to complete the purchase with a single click.

In Chapter 4, we discussed the design of an M-Business and topics related to the alliance value chain. M-Commerce transactions are good examples of how the alliance value chain is becoming more complex. A transaction may involve a customer, a wireless carrier, a wireless portal, content providers and merchants, and financial service providers. The number of roles present within these types of transactions means that all players in the value chain need to have seamless linkages between one another and be able to sort through the complex revenue sharing arrangements that may occur.

M-Commerce Platform Example

MoreMagic Software
http://www.moremagic.com

MoreMagic Software provides carrier-class, cross-platform payment solutions for mobile commerce. The company is headquartered in Newton, Massachusetts and was founded in 1997 in Helsinki, Finland.

The MoreMagic payment transaction platform is aimed at wireless carriers and wireless application service providers. It enables dynamic pricing and payment for any kind of network service or application. The platform can be accessed by a variety of end-user devices including mobile phones, PDAs, handheld computers, and the Internet. It supports a variety of wireless and Internet standards including WAP, I-Mode, GPRS, EDGE, CDMA, WDMA, and UMTS. The solution network architecture is illustrated in Figure 8–4.

Figure 8–4   MoreMagic Solution Architecture. Source: MoreMagic Software.

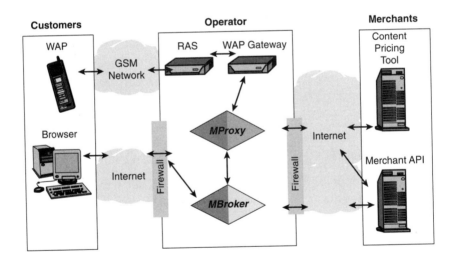

The platform is comprised of the following components:

>> **MBroker**—A transaction platform for mobile operators. It includes architectural components for business rules, a client connection conduit (CCC), payment adapters, and authentication adapters.

>> **MProxy**—An MBroker component that provides a turn-key payment solution for mobile operators and merchants. MProxy is located between the end user and the merchant. It implements the functionality for transparent transaction triggering.

>> **MWallet**—A prepaid and postpaid electronic wallet. For the end user, the MWallet provides access to a graphical view of account activity and profile information via Web or WAP interfaces.

>> **Merchant Tool Kit**—A set of documentation, code examples, and utility software to help merchants connect to an MBroker operator.

>> **Administrator Tool Kit**—Operation and maintenance software that enables the manipulation of the MBroker database and the monitoring of the MBroker behavior at run-time.

Considerations When choosing M-Commerce Platforms

Since M-Commerce business models are still very much formative, it is important when choosing a platform to ensure that the solution provides not only the technical flexibility to meet your current and future needs, but also the business flexibility. You should also look at the ease of use for both end users and for the other participants in the value chain, i.e., merchants, authentication service providers, and payment service providers. To help you in selecting your M-Commerce platform, here are some business and technical considerations.

Business Considerations

>> Does the M-Commerce platform provide support for the kinds of business models and revenue sharing arrangements that you plan to deliver?

>> Does the platform support the various kinds of product, service, or content that you plan to deliver?

>> Does the platform provide support for customizable business rules?

>> How you do plan to roll out your M-Commerce services? Can the platform vendor provide ways to reduce the costs involved for trial and pilot programs?

>> What are the software requirements, if any, for all the participants in the M-Commerce value chain? Are merchants required to install software or otherwise customize their sites in order to communicate with the M-Commerce platform operator?

>> What is the track record of the software company and its products?

>> How does the platform vendor bill for the services? Is it a standard license or a fee per transaction model?

Technical Considerations

>> Does the platform support the typical devices used by your customers?

>> Does the platform support the payment methods required by your customers?

>> Does the platform support integration with payment service provider systems?

>> Does the platform support integration with merchant systems?

>> Does the platform meet IT requirements in terms of scalability, performance, reliability, maintainability, security, manageability, and availability?

>> How well does the platform fit in with current IT standards and guidelines?

>> Does the platform support the provisioning, customer care, and billing functionality that is required?

>> How well does it support application management and systems management functions?

M-Business Implementation

Requirements Matrix and Vendor Selection

When a specific M-Business project opportunity has been identified in the business strategy, the initiative has passed the strategic opportunity assessment phase, funding has been secured, and the high-level project plan has been assembled, the next step is to start the implementation phase of the project. The project will usually follow a typical software development lifecycle such as requirements analysis, architecture and infrastructure planning, design, coding, testing, and implementation. Other techniques such as rapid prototyping are also commonly employed.

In this section, we take a look at how to make a vendor selection by use of a requirements matrix. This is a typical consulting approach that I have employed heavily in the past and takes into account the business and technical requirements and also their weightings in terms of relative importance in order to help facilitate the vendor selection process.

Vendor selections are typically made by the IT department, but for the business stakeholder this process is important to understand because business objectives need to drive the overall selection as much as the technical requirements.

Vendor selections can be made in as little as a couple of days or as much as several weeks, based upon the current knowledge of business and technical requirements, the knowledge of the various vendor technologies, and the newness of the category being evaluated.

The requirements matrix approach to vendor selection consists of the following phases:

>> Determine business requirements.

>> Determine technical requirements.

>> Assign relative weightings to requirements.

>> Determine candidate vendors and solutions.

>> Compile a requirements matrix in spreadsheet format.

>> Score vendors and solutions based upon strength in each requirements area.

Table 8–1 shows an example requirements matrix for a wireless middleware selection process. Vendor names have been excluded, but the table should serve to illustrate the technique and some of the requirements that may come into consideration. In this situation, the requirements were mostly of a technical nature related to the development and support of the applications.

Table 8-1 Example Requirements Matrix for Wireless Middleware Selection

Requirement	Weight-ing	Vendor A	Vendor B	Vendor C	Vendor D	Vendor E
End User Experience	3	8	8	8	8	8
Performance	5	8	8	6	8	8
Scalability	3	8	8	6	8	8
Reliability	5	8	8	6	8	8
Maintainability	3	8	6	8	6	6
Manageability	3	6	6	6	6	6
Security	5	8	8	8	8	8
Reuse/Components	3	10	8	6	8	8
Leverage of Existing Developer Skill Sets	1	6	6	10	8	6
Availability/Cost of Skilled Resources	3	6	6	10	8	6
Developer Learning Curve	1	6	6	10	6	6
Open Standards	3	8	8	6	8	8
Time to Market	5	8	6	10	6	6
Compliance/ Interoperability with Existing Apps and Tool Sets	5	6	6	8	8	6
Vendor Stability/ Maturity	3	6	10	10	10	8
Product Maturity/ Market Acceptance	3	6	8	8	8	8
Total Cost of Ownership	3	6	6	8	6	6
TOTAL		418	414	438	432	408

The best approach to capturing business and technical requirements for the solution is to have a meeting with stakeholders from both sides at the table: the business owners to state their requirements and expectations, and the IT owners to represent the development and support side of the equation. The best-fit solution must meet or exceed the business need, but must also meet the expectations of the IT department in terms of cost of ownership, ease of development, and support for existing standards. The top ten to twenty requirements should be captured and a weighting assigned to each requirement in order to detail the relative importance. A scale of 1 to 5 may be employed, with 5 denoting those requirements of highest importance.

Once the candidate vendors and solutions have been selected, each vendor solution can be assigned a score based upon its ability to meet the captured requirements. A scale of 1 to 10 may be employed with 10 denoting the highest score.

A total score for each vendor is compiled by multiplying the weightings and scores for each requirement and adding them together. There is nothing special about the numbers that come out of this scorecard. The numbers are simply a way to help determine which solution appears to have the best fit with the requirements relative to one another.

Of course, this process should be considered just an additional tool to help with vendor selection. One of the benefits to the approach is that the business users are given the opportunity to state which features and functions are of most importance to them and assign weightings. Also, the final scorecard presents a useful quantitative guide that can help to add to the qualitative data compiled regarding each candidate vendor solution.

In summary, the requirements matrix described above can be used as an additional tool during the selection of M-Business vendors and their products and services.

Implementation Strategies

M-Business implementations generally fall into two main categories: wireless enablement via wireless middleware and/or packaged applications, and transactional M-Commerce applications. Additionally, applications may be hosted within the enterprise or externally with wireless application service providers (WASPs). The following recommendations aim to provide some best practices in M-Business implementations. These best practices are similar to those of E-Business implementations, but include additional variables and risk factors such as the current diversity of devices, networks, and standards from which to choose.

>> **Determine Whether Applications Should be Outsourced or Hosted Internally**—One of the first decisions to be made is whether to outsource the wireless application initiatives or whether to bring them in-house. This will be a matter of how strategic the applications are to your business and the level of internal competency or desire to implement and support such applications.

>> **Pursue a Phased Rollout Approach in Terms of User Constituencies**—To manage risk, pursue a phased rollout of your applications. You may be able to start with IT staff and then roll out applications to corporate users and eventually to outside customers. This is very much application dependent (based on the target users), but in general this will provide an opportunity to gain valuable feedback on the application usability and support procedures prior to full rollout.

>> **Rollout Application Functionality in Phases**—Some of the most successful M-Business applications deployed by early adopters have been deployed in small functional increments with simple applications such as order status as initial features. It is important to gain user acceptance with these initial features before expanding the application functionality available to end users.

>> **Leverage Existing User Devices**—To conserve costs, you may well be able to leverage existing devices owned by employees and customers in your application rollout. This opportunity to conserve costs by deploying to existing devices should be carefully balanced with the need to standardize on a common device for ease of support and maintenance moving forward. Again, this will be application-dependent based on the target user constituencies whether internal or external. For outside customers and business partners, you may have no option but to support the wide variety of devices in use such as pagers, SMS, and WAP-capable cell phones and

PDAs. For employees, you will need to determine whether to support the existing personally-owned devices (such as variety of Palm Pilots and Compaq iPAQs) or to invest in providing company devices in order to reduce support costs.

>> **Test Applications and Devices Thoroughly**—With M-Business technologies still maturing, enterprise caution surrounding the devices, networks, and standards is very appropriate. It is highly important to test these applications both from a software functionality perspective and a general usability perspective in real-life scenarios. Devices often have glitches that can impact their functionality. Operating systems on PDAs are not as robust as PC operating systems and battery life can also be an issue. Be sure to test the application functionality and usability in terms of navigation and user experience, but also test the usability of all devices that you are supporting. If devices have known issues be sure to document the support procedures so that users can be up and running again very quickly.

>> **Look for Unexpected Business Benefits**—With many of the case studies profiled in this book came unexpected business benefits. Be sure to capture feedback from end users on how they are using their wireless applications and any surprise uses that have emerged. This feedback should be incorporated into plans for future functionality.

>> **Evaluate Future Connectivity Options**—As some of the case study enterprises were rolling out their wireless applications, they were also exploring future connectivity options. Be sure to evaluate the full range of connectivity options such as cradle, wireless LAN, and wireless WAN. Even if some options are not yet ready or suitable for deployment, the lessons learned will prepare you to quickly move to these options as the technology matures.

>> **Evaluate Future Device Options**—In a similar manner to evaluating future connectivity options, be sure to evaluate future device options. Increased functionality in some of the newer devices may mean that you can reduce the number of devices supported from perhaps three or more to just two devices.

Business Agility Lessons

M-Business Technologies

>> Application servers, enterprise application integration (EAI) technologies, and wireless middleware will converge toward a common platform within the next 12 to 24 months.

>> Ecosystems of partnerships are already being assembled.

>> Software vendors are approaching M-Business from many different angles, including the packaged application perspective, the infrastructure perspective, and the specialty wireless application perspective.

>> When selecting wireless middleware platforms, determine how this technology fits into your overall IT strategy for M-Business.

>> When selecting M-Commerce platforms, due to the formative nature of the business models, it is important to look for business rule flexibility in addition to technical flexibility.

M-Business Implementation

>> A requirements matrix can help facilitate the vendor selection process by taking into account the business and technical requirements and also their weightings in terms of relative importance.

>> Some guidelines when implementing M-Business applications include the following:

  – Determine whether applications should be outsourced or hosted internally

  – Pursue a phased rollout approach in terms of user constituencies

  – Rollout application functionality in phases

  – Leverage existing user devices

  – Test applications and devices thoroughly

  – Look for unexpected business benefits

  – Evaluate future connectivity options

  – Evaluate future device options

9

Future Trends

W hen we think about and explore future trends for M-Business within an enterprise, we need to look at the entire value chain including devices, networks, standards, and applications. This gives us insight into some of the emerging technologies that are forming on the supply side of the equation and how their combination may create totally new opportunities for business agility to be achieved and enterprise business value to be extracted.

On the demand side, which is the real driver behind these future trends, we will continue to see business drivers of cost reduction, revenue generation, customer satisfaction, increased productivity, reduced cycle times, and so forth. It will be business as usual in terms of these corporate objectives, but we will see an increased focus on achieving business agility and the accompanying attributes of speed and flexibility that we discussed in the opening chapter of this book. Innovation via emerging technologies will always be a mechanism for pioneers and early adopters to differentiate themselves from their competition and gain some of these benefits. Today's emerging technologies present us with a unique opportunity to achieve levels of business agility that were previously unimaginable. Advances in processing power, network bandwidth, and application functionality mean that the enterprise can move toward a real-time computing paradigm where decisions can be made based upon the latest enterprise events.

General Themes Toward Business Agility

I think the general theme for business agility within the enterprise will be the continued automation of business processes via emerging technology and communications. We will have access to information and transactions at any time and any place. Additionally these applications will gain intelligence and become more aware of our context, both in terms of our role and our surrounding environment at any specific place and time. This role awareness may include awareness of the business activity we are performing, the physical objects around us, and the people with whom we are interacting. The technologies will, in effect, create an intelligent window into our business activities, both in terms of our real world and virtual surroundings. Tomorrow's applications will know a lot more about your current physical environment and activities. They will also know about your virtual business environment and activities beyond your current physical location.

The technology that automates us will also become better at blending into our process. The best technologies will be almost unnoticeable as we use them and will be nondisruptive to our activities. They will move with us as we move. They will seamlessly transition from personal-area network, to local-area network, to wide-area network. They will transition from online to offline modes and from one device to another. They will be "always on" technologies, which are ready to serve whenever and wherever needed. They will provide us with the most efficient interfaces for interaction based upon our environmental context—seamlessly switching between voice and data access as appropriate. They will learn our preferences, give us the most appropriate and meaningful content at all times, but always allow us to change our preferences and views into this information whenever we choose.

In this final chapter, we'll take a look at some of the future scenarios for the M-Business networks, devices, standards, and applications that will start to shape the next several years for consumers and businesses alike.

Networks

Perhaps the first thing to come to mind when someone talks about the future of wireless applications is the network. Third generation (3G) networks have been much publicized and discussed. When they arrive, they promise to bring a wealth of media-rich applications with them. In particular, they promise to include streaming audio and video and the ability for large file transfers together with "always-on" functionality.

Networks with high bandwidth and good coverage will certainly help move M-Business along its continuum. As the technology matures, enterprises will be able to move from wired, batch-like synchronization techniques that support mobile computing initiatives toward more real-time, wireless synchronization, and eventually to an always-on scenario where connectivity to enterprise systems is always available. Always-on connectivity will also enable real-time computing with information able to pass from source to destination freely and rapidly—enabling improved enterprise operations and decision making.

This always-on functionality will certainly have an effect upon the need for synchronization between mobile devices and server-side enterprise applications. Because of this, it will also have an effect upon the need to store data on the client device. In reality, this end point of constant connectivity may be hard to achieve owing to the geographic irregularities in the physical world that make 100% wireless coverage a challenge owing to buildings, hills, and other types of interference. But with a patchwork of wireless local-area networks and wireless wide-area networks, we will be able to achieve tremendous improvements in bandwidth and accessibility to information.

In the short- to mid-term, enterprises may need to continue to embrace offline techniques on their mobile devices as a way to continue business operations, such as field force and sales force automation when wireless connectivity is unavailable. In the longer term, always on wireless connectivity may make this less of a necessity.

What is likely to evolve in the wireless network space over the next few years is a patchwork of wireless connectivity options based upon the location of the user. Bluetooth networks will likely take hold in personal-area network settings of public spaces and retail stores, where devices can communicate with Internet access points, cash registers, vending machines, and other personal devices. Wireless LAN networks, such as those supported by the 802.11 standards, will likely take hold in more corporate settings such as corporate office complexes and warehouses as a way to get onto the corporate intranet and to access corporate applications such as enterprise resource planning systems and warehouse management systems. Finally, wireless networks such as the 3G networks provided by the carriers will likely take hold in outdoor scenarios, where the field force and sales force need access to corporate applications as part of their regular work activity. Thus, there may be no single network technology that wins in all areas of enterprise activity. The winners may likely be spread across those supporting the Bluetooth standard, the wireless LAN standards such as 802.11, and the 3G standards of the wireless carriers.

Another growth area will be that of cross-network roaming and interoperability, and intelligent matching of content to wireless transport. Innovative companies will spring up in order to help alleviate the challenges that consumers and enterprises may face when moving from one network boundary to the next, and the challenges that carriers face in managing their spectrum efficiency, bandwidth, and content. As an example, Cyneta Networks based in Plano, Texas, is creating a resource aware adaptive content switch that optimizes IP communication in the wireless environment. This allows operators to provide differentiated levels of Quality-of-Service (QoS), enhanced session management, performance measurement, and more reliable transport for wireless data traveling over their networks.

Devices

Devices are certainly changing forms right now, as manufacturers experiment with device form factors and attempt to increase their usability and functionality for various enterprise and consumer scenarios.

Wearable devices may be a future trend as mobile devices morph into the ultimate in mobility by becoming wearable as gloves, headsets, and so forth. Wearable devices benefit the end user by being located at the point of activity, but not interfering with the activity at hand. They can provide improved interactivity and flexibility over the traditional mobile devices or the traditional keyboard and monitor. Essentially, they provide the advantages of automation without the disadvantage of the process change to accommodate the technology.

Another future trend that is readily apparent today is the combined cellular phone and PDA—often termed the smartphone. Devices such as Nokia's 9290 Communicator are fully integrated mobile terminals running the Symbian operating system that combine phone, fax, e-mail, calendar, and imaging functionality into a single device. Kyocera's QCP 6035 smartphone combines a CDMA digital wireless phone with a Palm OS handheld computer. Ericsson's R380 is another example of a smartphone having PDA functionality and running on GSM networks. Beyond the cell phone and PDA combination, companies such as Samsung have introduced concept phones including watch phones, camera phones, and TV phones.

Devices such as the smartphone may well enable the enterprise to standardize on fewer devices for their corporate users. Fewer devices translates into reduced purchase and support costs and increased productivity for employees.

Wearable Devices

SAP provides a glimpse into this future world of wearable devices in their mobile future demo, which is part of the mySAP Mobile Business section of their Web site. The scenario is described in Table 9–1.

According to SAP, this scenario provides an example of "mobile, hands-free interaction—a new way to compute." The benefits include the elimination of printed lists and improved ability to locate packages based upon the unique functionality of the wearable equipment, which can automatically recognize packages and inform users as to their contents.

Table 9–1   SAP "Mobile Future" Scenario (Source: SAP)

Scenario

"Pat Miller is the Quality Manager within the goods issuing department at yournet.com. He checks parcels before they are delivered to the customer. He doesn't have access to a desktop computer. He is wearing his computer...and monitors the processing with his private eye headset. He picks up the package and his special glove accesses the corresponding data. The system recognizes the package and shows him what it should contain. It's all in the palm of his hand."

Benefits

>> Information by contact

>> Automated processing and interaction

>> Includes frontline workers in E-Business processes

>> Hands-free operation

Possibilities

>> Proof of delivery for inbound processing

>> Loading control for outbound processes

>> Quality inspection

>> Inventory management

This example shows how wearable devices can benefit field workers within the enterprise by giving them tools that provide additional information needed to perform their jobs more effectively.

An example of a company that provides products in the wearable computing space is Xybernaut Corporation, based out of Fairfax, Virginia. The company has a product named the Mobile Assistant V that is a wearable computer that comprises a CPU module, a head-mounted display or flat-panel display, and a mini-keyboard. The solution supports various operating systems including Windows 98, Windows NT, Windows 2000, Linux, and Unix, and is powered by an Intel 500 MHz processor designed for ultra low voltage. Specialized optics on the head-mounted display which measures just over an inch in diameter provide the same full-screen, color view as a standard 15" monitor. The solution supports wireless connectivity via wireless LAN or dial-up cellular connections. These wearable solutions have particular relevance in a number of enterprise scenarios including installation, communications, inspection and maintenance, inventory and training. Sample customers using the Xybernaut solution include BOC Gases, Geophysical Survey Systems, Inc., FedEx Corporation, Bell Canada, and the U.S. Army and Navy.

Smart Phones

Smart phones provide a number of benefits for the enterprise and for consumers alike. As mentioned earlier, one of the main benefits for the enterprise is the fact that they may help enterprise IT departments to standardize on fewer devices for their corporate users. Instead of a corporate user having to carry around a pager, cell phone, PDA, and laptop, they may eventually be able to carry around just two devices: a laptop and a smartphone. The smartphone will effectively perform all the functions of the previous pager, cell phone, and PDA devices. Which vendors will ultimately be the most successful in this battle for next generation device supremacy is still to be determined. The vendors are approaching it from a number of different angles including adding PDA features to cell phones and vice versa.

An example of the PDA with added cell phone functionality is the Handspring VisorPhone. The VisorPhone is a Springboard expansion module that can be attached to any standard Handspring Visor PDA in order to enable cellular calls, wireless Internet access, and text messaging.

Examples of cell phones with added PDA functionality include the Nokia 9290 Communicator, Kyocera QCP 6035, Ericsson R380, and several others. Several of these smartphones run the Symbian operating system. Symbian is a privately held company with offices in the UK, Sweden, Japan, and the U.S. It was established in June 1998 by Ericsson, Motorola, Nokia, and Psion. Matsushita (Panasonic) joined as a shareholder in May 1999. The companies' mission statement is to set the standard for mobile wireless operating systems and to enable a mass market for Wireless Information Devices.

Standards

The emerging standards that will shape the M-Business arena in the future may well be a combination of both wireless and non-wireless specific standards. In fact, M-Business applications within the enterprise may come to be shaped by an equal portion of both categories of standards. Certain wireless-specific standards such as location-based services, synchronization standards, voice standards, and device operating system and browser standards will obviously play an important role, but also key developments in more mainstream Internet standards will play an important role as well.

Some of the wireless-specific and wireless-oriented standards include the Java 2 Platform, Micro Edition (J2ME), Binary Runtime Environment for Wireless (BREW), the Wireless Application Protocol (WAP), Voice eXtensible Markup Language (VoiceXML), and SyncML.

The Internet standards include XHTML and the work that is being conducted on the semantic Web. Other standards that will likely impact M-Business applications in the future include the numerous Web services and enterprise application integration-oriented standards such as ebXML, Java Message Service (JMS), Simple Object Access Protocol (SOAP), Universal Description, Discovery and Integration (UDDI), Web Services Description Language (WSDL), and standards within software categories such as business process management, natural language processing, and real-time computing.

>> **Java 2 Platform, Micro Edition (J2ME)**—The J2ME technology is Sun's version of its Java technology for the consumer space and networked device market. It includes the range of devices from smart cards to pagers and cell phones to set-top boxes. An example of a cell phone currently running the J2ME platform is the i85s from Motorola with service available from Nextel. The J2ME platform allows devices to download and receive not only new applications, but also new libraries that form part of the J2ME platform itself.

>> **Binary Runtime Environment for Wireless (BREW)**—The BREW platform allows developers to create applications that operate on handsets with Qualcomm CDMA chipsets. BREW sits between

the chip system software and the applications and enables developers to gain access to the phone functionality via a Windows-based software development kit. Using BREW, developers can build handset applications that users can download over carrier networks onto BREW-enabled phones.

>> **Wireless Application Protocol (WAP)**—Version 2.0 of the Wireless Application Protocol will include support for XHTML, TCP, color graphics, animation, large file downloads, location-aware services, pop-up and context-sensitive menus, and data synchronization with desktop personal information management systems. In the security area, it supports Public Key Infrastructure (PKI) and brings the level of security up to the level of wired Web sites by supporting end-to-end encryption and allowing for secure proxies in handsets and gateways.

>> **Voice eXtensible Markup Language (VoiceXML)**—VoiceXML is an XML-based specification defined by the VoiceXML forum that will help to open up proprietary telephony platforms and enable an enterprise to gain voice access to their data and applications. One of the companies developing products in this space is Informio—an enterprise voice solutions provider. Some of the applications of voice access within the enterprise include the ability to hear e-mail messages while on the road, to record a voice message as a reply to an e-mail, and to forward an e-mail to someone in your directory.

>> **SyncML**—SyncML is an open industry standard based upon XML for the universal synchronization of remote data and personal information across multiple networks, platforms, and devices. The initiative is sponsored by Ericsson, IBM, Lotus, Matsushita, Motorola, Nokia, Openwave, Starfish Software, and Symbian. The SyncML initiative maintains a Web site at http://www. syncml.org, which provides information and an on-going list of compliant client and server products.

>> **eXtensible Hypertext Markup Language (XHTML)**—XHTML is a reformulation of HTML 4.0 as an XML 1.0 application. The specification is currently a W3C recommendation. In addition to moving HTML into the XML world with all of its benefits, the specification is also enabling HTML to better support the wide

variety of access devices that are emerging beyond the desktop PC and browser. XHTML provides an extension and subsetting mechanism so that content can be developed once and targeted for different devices. XHTML is an evolutionary step in the future of Internet standards and will help extend the reach of these standards beyond the traditional desktop.

>> **Web Services and Enterprise Application Integration**—EAI standards such as ebXML, JMS, SOAP, UDDI, and WSDL will impact M-Business applications and business agility in general owing to their ability to ease the integration process and communication process between disparate applications, whether intra-enterprise or inter-enterprise. As wireless and mobile applications become more strategic and tie in to greater numbers of enterprise back-end applications, these emerging standards will help ease the implementation process and ensure that the wireless channel has the same access to information as the wired channel.

All of these standards outlined above will help advance the possibilities for M-Business by increasing functionality, accessibility, interoperability, security, and general usability. They will help to lay the groundwork toward increased adoption of wireless and mobile applications and the eventual mobile information society.

Applications

Future applications that are deployed in mobile scenarios will leverage and build upon the innovations in the networks, in the devices, in the device operating systems and browsers, and in Internet standards in general. Such innovations will include 3G networks, smartphones such as the Nokia 9290 Communicator, new wireless internet device operating systems such as Microsoft's Stinger and the Symbian OS, new browser technologies such as Microsoft's Mobile Internet Explorer and Openwave's Mobile Browser, and emerging or maturing wireless and Internet standards such as XHTML, WAP, J2ME, BREW, SyncML, VoiceXML, and many others. These are just a few of the technologies and standards that are evolving rapidly and will help to improve the end-user experience for M-Business and the development and deployment of these applications.

A company that is working to integrate and prioritize applications and content into a coherent summary for delivery through wireless voice and data services is Sirenic, based in Mountain View, California. The company has a five-step process for delivering appropriate content consisting of the following steps: categorize, prioritize, personalize, stratify, and render. Companies like Sirenic are helping deliver relevant content to wireless appliances, interactive TV systems, and CarPC systems and are minimizing the required number of interactions to perform a given task.

Some of the categories of software application within the enterprise that will impact M-Business and overall business agility include business process management, natural language processing, real-time computing, the semantic Web, and Web services.

>> **Business Process Management (BPM)**—Business process management is a category of software that aids in the development and operation of business processes across multiple applications and business partners. BPMI.org is a non-profit organization that aims to promote and develop the use of BPM within the industry by establishing open standards and specifications. BPM can be considered a driver or superset of enterprise application integration where the integration is orchestrated by a governing process model. The EAI integration points are public and private connection points in the overall business process flow. M-Business plays a role in the overall process where human-interaction is required for approvals and other process steps along the way.

>> **Natural Language Processing (NLP)**—Natural language processing involves the development of algorithms that can better "understand" the meaning of a human question when spoken or written naturally. Rather than the searches that are most common on the Web today, such as keyword searches which have unpredictable and often irrelevant results, NLP can determine the semantics of the question, often in a variety of languages, and provide a more concise and relevant answer based upon the scope of its searchable index. The technology that I have seen both in the academic world and the corporate world is extremely accurate and can answer highly specific questions with exact answers—often down to a couple of words that provide the required answer. Production implementations of natural language processing

include the English Query functionality within Microsoft's SQL Server database management system. Instead of having to write complex SQL statements in order to perform a database query, users can use the English query interface in order to perform their queries using a naturally written question. NLP is an excellent example of technology adapting to our natural communication mechanisms and developing greater understanding of the meaning of those communications.

>> **Real-Time Computing**—Real-time computing is another area that speaks to business agility. The concept involves speeding the reaction time of the enterprise for a variety of business processes. If the enterprise can react more quickly to internal and external stimuli, it can make quicker decisions and, if necessary, change its behavior to reduce costs or improve revenues. An example is the supply chain. If the enterprise can see inventory buildup in real-time or close to real-time, it can act more quickly to reduce production and hence avoid or minimize inventory write-offs. Real-time computing is closely linked with M-Business because in order to achieve real-time computing one needs access to information inputs, outputs, and decision-makers regardless of location. For example, if data can be captured by a field worker as soon as an inspection is performed, then that information can immediately be relayed to a decision maker or approver for the appropriate action to be taken within seconds of the initial input. This stands in stark contrast to some of today's processes, where data is often recorded manually in the field, input into an enterprise application days or weeks later by a data-entry clerk, and not acted upon for weeks or even months later.

>> **Semantic Web**—Just as it is important for a machine to be able to understand a human as in the natural language processing topic that we discussed, it is also important for machines to be able to understand other machines. The semantic Web aims to increase the meaning associated with information located on the Internet. Today the information is readily understandable by humans, but to a machine it is less so. By adding meaning to the information on the Web, the semantic Web vision can enable machines to interop-

erate and collaborate with one another to perform many beneficial tasks. Intelligent software agents will be able to visit many sites and compile information and perform transactions on our behalf. They will be able to interpret the information that they find on these sites and make intelligent decisions on our behalf. They will also be able to confirm with us those decisions that are a matter of personal preference so that we are able to maintain oversight on these tasks and activities as they occur. The semantic Web activity is being conducted by the W3C and is being developed by leveraging XML and the Resource Description Framework (RDF). RDF is a mechanism for metadata representation and transport and as such allows developers to attach named properties and property values to the resources they are describing.

>> **Web Services**—As we noted in the standards section, Web services are a way for companies to expose the business services they provide over the Internet for others to discover and utilize. This speeds business agility by lowering the time required to integrate with a business partner and use their services or to perform transactions. It also ties back to our concept of the alliance value chain where business relationships can be located, created, transacted, and managed as part of the alliance life cycle with alliance chain management and alliance chain integration being two important aspects from a business and technical viewpoint, respectively. While technology can be used to help speed the discovery, integration and ongoing measurement of dynamic business relationships, the business issues behind these relationships will always require at least some level of human interaction so collaborative platforms in addition to integration platforms are highly desirable.

The five software categories mentioned above represent just a sampling of some of the upcoming applications and services that promise to shape the future for M-Business and overall business agility. In the conclusion, we'll summarize the trends in devices, networks, standards, and applications in order to see how all these innovations are shaping our information technology future and the realm of possibilities for the enterprise.

Business Agility Lessons

Networks

>> Always on networks will provide the potential for the real-time enterprise.

>> Seamless transition from personal-area network to local-area network to wide-area network.

Devices

>> Smartphones may help to reduce the number of devices that enterprises have to support.

>> Wearable devices will help to make M-Business applications non-invasive. They provide the advantages of automation without the disadvantage of the process change to accommodate the technology.

Standards

>> The emerging standards that will shape the M-Business arena in the future may well be a combination of both wireless and non-wireless specific standards.

Applications

>> Applications that will increase overall business agility include business process management, natural language processing, real-time computing, the semantic Web, and Web services.

>> Attributes of emerging applications will include:

- Information and transactions anytime and anyplace

- Intelligent, context-aware applications

- Non-disruptive applications that blend into work activities and processes
- Seamless transition from online to offline modes

- Seamless transition from one device to another

- Seamless switching between voice and data access for efficiency

Conclusion

In coming to the end of our exploration of M-Business within the enterprise, several interesting themes have emerged. The mixture of case studies that we have profiled reveals some interesting data points about where we are on the M-Business continuum. For all of these companies covered in the case studies, enterprise value is being enhanced through M-Business technologies. The difference is that the value is enhanced through different technology leverage points. All the solutions have included a high degree of mobility, and high degree of process automation, but they have adopted varying degrees of wireless connectivity. Some have adopted full wireless connectivity, such as the wireless LANs within hospitals. Others have adopted part-time wireless connectivity, and others have adopted wired connectivity via nightly batch synchronization.

Since wireless is still an emerging and maturing industry in terms of enterprise adoption of wireless data and applications, what is happening is that enterprises are tuning their level of wireless adoption based upon their level of need for business agility. Those early adopters who need the highest levels of business agility are pushing the envelope and adopting fully wireless solutions in order to meet their needs for real-time access to information and transactions regardless of location. Those enterprises that have slightly lower requirements for business agility, at least at the present time, are adopting mobile solutions or partial wireless solutions in order to meet their needs for perhaps daily access to information and transactions when mobile.

The trend here is that the enterprise is requiring faster and faster access to information. We have more information that is being collected and exchanged across all areas of the enterprise and all user constituencies. This information is changing more rapidly as time progresses, and we need more real-time access to this information.

Just as the desktop computer has started to become a thing of the past with more and more enterprises equipping their staff with laptops, so we will see more enterprises equipping their staff with devices that are increasingly mobile and increasingly connected in order to speed business agility.

The macro trend that is occurring when we look at M-Business initiatives within the enterprise and also within our daily lives as consumers, is that we are finally shaping technology around ourselves as opposed to shaping ourselves around technology. As we do this, technology is gaining more human-like attributes as follows:

>> **Context-Aware**—Just as we are aware of our surroundings at all times, so our M-Business applications are becoming aware of our location and our current activities. For example, once location-based services become widespread, location information can be used to determine the user's context of activity such as home or office.

>> **Always-On**—Just as our senses are continually providing input for us to react to, so our M-Business applications will become always-on, effectively plugging us into our corporate networks including both people and machines. For example, just as RIM pagers give us always-on access to e-mail today, so will the 3G wireless networks provide us with richer forms of always-on functionality such as access to enterprise applications in the future.

>> **Multi-Sensory**—Just as we leverage many senses in order to communicate, so our M-Business applications will support many modes of communication, including voice and data. They will seamlessly switch between the two. For example, speech may be the most appropriate and safe communication mechanism for navigating the Web or requesting information while in transit. Standards such as VoiceXML and SyncML will help drive device and voice/data interoperability.

>> **Non-Invasive**—Just as we go about our day-to-day activities without having to think about the mechanics of how we move or how we think, our M-Business applications will become non-invasive to our work activities. They will provide us with the information we need, but without requiring us to adapt around these devices. For example, wearable devices will enable us to minimize footsteps and capture information at the point of activity.

>> **Interpretive**—Just as we are able to interpret the information that our senses pick up, so our M-Business applications will be able to interpret information on our behalf and be able to provide us with the most relevant information. For example, the semantic Web and natural language processing will enable computer applications to better understand the meaning of content and make smarter decisions on our behalf.

>> **Intelligent**—Just as we are able to apply intelligence and reasoning before making decisions, so our M-Business applications will gain intelligence and relevance by being tied into business process management scenarios, rules-based engines, and workflows. Business intelligence applications will help us get the right information related to key business metrics and competitive data to the right people at the time when it is most needed.

>> **Responsive**—Just as we are able to react in real-time to external stimuli, so our M-Business applications will provide us with alerts to key business events in real-time for improved enterprise decision making. Real-time computing will be a software category that helps to drive this forward, along with the mobility and always-on functionality of our wireless applications and networks.

>> **Adaptive**—Just as we are able to adapt to our surroundings, so our M-Business applications will be able to adapt to our enterprise context. This may include adaptation to the inputs that are collected, to the processing that is performed, and to the outputs that are distributed. Standards such as WAP 2.0, J2ME, and BREW will help devices become more usable and adaptive to our needs in the future.

Going back to the formula that we introduced at the beginning of the book, we stated that business agility was the sum of process agility and technical agility. We also stated that mobile business was the sum of business process, electronic business, and wireless communications as follows:

> *Mobile Business = Business Process + Electronic Business + Wireless Communications*

The best way to extract enterprise value from mobile business is to leverage it from the standpoint of all three of these disciplines. Mobile business should be viewed as a superset of the three and not a mere intersection of wireless communications and electronic business, as has often been proposed in the past. It should also be viewed in the broader context of business agility.

We are well on the way toward adapting technology around our business needs and activities. We are also shaping technology and our business operations around ourselves by adding human-like characteristics. Our enterprise operations and activities are becoming faster, more flexible, more mobile, and more intelligent. Today, mobile business is a reality. Tomorrow, true wireless business will also become a reality and the enterprise will be one step further towards the ultimate goal of business agility.

Business Agility Lessons

>> Enterprises are tuning their level of wireless adoption based upon their level of need for business agility.

>> We are finally shaping technology around ourselves as opposed to shaping ourselves around technology.

>> M-Business technologies and applications are gaining human-like attributes, including the ability to exhibit the following: context-aware, always-on, multi-sensory, non-invasive, interpretive, intelligent, responsive, and adaptive.

Diagnosing Your Business Agility

>> What is your current level of business agility and how rapidly can you respond to change?

>> What is the level of business agility of your competitors?

>> What information, transactions, and decisions can you pull or push out to the edges of your organization for increased business agility?

>> What emerging technologies can you apply to enhance your business agility?

Notes and References

Chapter 1

1. Figure 1-2: *The Ten Day MBA Revised Edition*, Steven Silbiger, page 299. "Reprinted from Business Horizons, June 1980. © 1980 by the Foundation for the School of Business at Indiana University. Used with permission."

2. Figure 1-4: *Business @ The Speed of Thought*, Bill Gates, page 118. "Source: W. Michael Fox and Forbes Magazine."

3. Figure 1-5: *Hype Cycle of Emerging Technology*, Gartner Website Glossary, July 2001, http://btj.gartner.com/btj/static/glossary/glossary_h.html.

4. Figure 1-7: VentureOne, Q1 Liquidity Data (liquidity1Q01.xls), http://www.ventureone.com

5. In an online poll conducted by *Internet Week*, 86% of respondents stated that they believed both IT and business management should be involved in IT decisions.

Chapter 2

6. Figure 2-1: Internet User Penetration By Region. Source: ARC Group, "Content & Applications for the Wireless Internet—Worldwide Market Analysis & Strategic Outlook 2000-2005," Table 3.1.

7. Figure 2-2: Mobile Data User Penetration By Region. Source: ARC Group, "Content & Applications for the Wireless Internet—Worldwide Market Analysis & Strategic Outlook 2000-2005," Table 3.2.

8. According to the research firm Jupiter, there will be one billion wireless Web devices in circulation by the year 2003. They also go on to say that companies must enable wireless extensions during the next 12-18 months, or risk losing customers to competitors that do. Source: Jupiter.

9. Table 2-1: Analyst Predictions for Wireless Data and M-Commerce.

10. Figure 2-3: Comparison of 1G, 2G, 2.5G, and 3G Networks. Adapted from Nokia (http://www.nokia.com/3g/whatis.html) and 3G Newsroom (http://www.3gnewsroom.com/html/what_is_3g/index.shtml).

11. GSM Association, 25 May 2001, "50 Billion Global Text Messages in Q1!," http://www.gsmworld.com/news/press_2001/press_releases_22.html

12. Safeway, http://www.safeway.co.uk

13. NTT DoCoMo, http://www.nttdocomo.com

14. Table 2-2: Wireless Internet IT Concerns. Source: *Internet Week*, "Wireless Slow To Catch On," December 11, 2000, Bob Violino and John Webster.

15. According to *Strategy Analytics*, the market for location-based services will reach $6.5 billion in the United States and $9 billion in Europe by 2005. Source: Strategy Analytics.

16. According to Nokia, over 80 3G licenses were granted in 2000, and several hundred more will be granted over the next few years. Source: Nokia, http://www.nokia.com.

17. According to 3G Newsroom (http://www.3gnewsroom.com), the five most expensive auctions were Germany, Britain, USA, Italy, and South Korea with a total of over $100 billion spent within these five countries alone. Source: 3G Newsroom, http://www.3gnewsroom.com.

18. Figure 2-5: Predictions for Increase in Value-Added Data Services in Order to Raise Carrier ARPU. Source: ARC Group.

19. Giga Information Group, "Top-Ten Problems that Customers Have with Their IT Organizations," Marc Cecere, May 10, 2001.

20. Table 2-4: M-Business 50 Stock Index. Source: *M-Business Magazine*, http://www.mbizcentral.com

Chapter 3

21. *Forbes, Special Issue: Strategic Alliance Guide*, May 21, 2001.

Chapter 4

22. SiteStuff, http://www.sitestuff.com, "Facility Management Goes Wireless," 5/27/2000; Notifact.com, http://www.notifact.com; USTelemetry.com, http://www.ustelemetry.com

Chapter 7

23. Rational, http://www.rational.com

Chapter 9

24. SAP—mySAP Mobile Business—http://www.sap.com/solutions/mobilebusiness/

25. Xybernaut, http://www.xybernaut.com and http://www.zdnet.com/eweek/stories/general/0,11011,2715691,00.html

Index

The *Financial Times* delivers a world of business news.

Use the Risk-Free Trial Voucher below!

To stay ahead in today's business world you need to be well-informed on a daily basis. And not just on the national level. You need a news source that closely monitors the entire world of business, and then delivers it in a concise, quick-read format.

With the *Financial Times* you get the major stories from every region of the world. Reports found nowhere else. You get business, management, politics, economics, technology and more.

Now you can try the *Financial Times* for 4 weeks, absolutely risk free. And better yet, if you wish to continue receiving the *Financial Times* you'll get great savings off the regular subscription rate. Just use the voucher below.

4 Week Risk-Free Trial Voucher

Yes! Please send me the *Financial Times* for 4 weeks (Monday through Saturday) Risk-Free, and details of special subscription rates in my country.

Name _____

Company _____

Address _____ ❏ Business or ❏ Home Address

Apt./Suite/Floor _____City _____State/Province_____

Zip/Postal Code_____Country _____

Phone (optional) _____E-mail (optional)_____

Limited time offer good for new subscribers in FT delivery areas only.

To order contact Financial Times Customer Service in your area (mention offer SAB01A).

The Americas: Tel 800-628-8088 Fax 845-566-8220 E-mail: uscirculation@ft.com

Europe: Tel 44 20 7873 4200 Fax 44 20 7873 3428 E-mail: fte.subs@ft.com

Japan: Tel 0120 341-468 Fax 0120 593-146 E-mail: circulation.fttokyo@ft.com

Korea: E-mail: sungho.yang@ft.com

S.E. Asia: Tel 852 2905 5555 Fax 852 2905 5590 E-mail: subseasia@ft.com

www.ft.com

FT FINANCIAL TIMES
World business newspaper

8 reasons why you should read the Financial Times for 4 weeks RISK-FREE!

To help you stay current with significant
developments in the world economy ...
and to assist you to make informed business
decisions — the Financial Times brings you:

1 Fast, meaningful overviews of international affairs ... plus daily
briefings on major world news.

2 Perceptive coverage of economic, business, financial and political
developments with special focus on emerging markets.

3 More international business news than any other publication.

4 Sophisticated financial analysis and commentary on world market
activity plus stock quotes from over 30 countries.

5 Reports on international companies and a section on global investing.

6 Specialized pages on management, marketing, advertising and
technological innovations from all parts of the world.

7 Highly valued single-topic special reports (over 200 annually)
on countries, industries, investment opportunities, technology and more.

8 The Saturday Weekend FT section — a globetrotter's guide to
leisure-time activities around the world: the arts, fine dining, travel,
sports and more.

For Special Offer See Over

FT FINANCIAL TIMES
World business newspaper

Where to find tomorrow's best business and technology ideas. TODAY.

- Ideas for defining tomorrow's competitive strategies — and executing them.

- Ideas that reflect a profound understanding of today's global business realities.

- Ideas that will help you achieve unprecedented customer and enterprise value.

- Ideas that illuminate the powerful new connections between business and technology.

ONE PUBLISHER.

Financial Times
Prentice Hall.

FINANCIAL TIMES
Prentice Hall

WORLD BUSINESS PUBLISHER

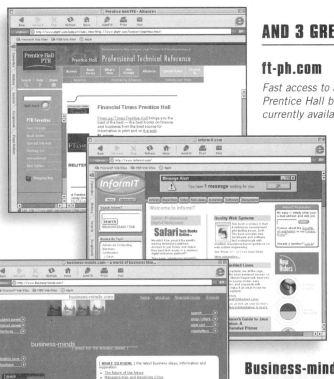

AND 3 GREAT WEB SITES:

ft-ph.com

Fast access to all Financial Times Prentice Hall business books currently available.

InformIt.com

Your link to today's top business and technology experts: new content, practical solutions, and the world's best online training.

Business-minds.com

Where the thought leaders of the business world gather to share key ideas, techniques, resources — and inspiration.